Science and Technology in the Management of Complex Problems

ORGANISATION
FOR ECONOMIC CO-OPERATION
AND DEVELOPMENT

The Organisation for Economic Co-operation and Development (OECD) was set up under a Convention signed in Paris on 14th December, 1960, which provides that the OECD shall promote policies designed:

— to achieve the highest sustainable economic growth and employment and a rising standard of living in Member countries, while maintaining financial stability, and thus to contribute to the development of the world economy;
— to contribute to sound economic expansion in Member as well as non-member countries in the process of economic development;
— to contribute to the expansion of world trade on a multilateral, non-discriminatory basis in accordance with international obligations.

The Members of OECD are Australia, Austria, Belgium, Canada, Denmark, Finland, France, the Federal Republic of Germany, Greece, Iceland, Ireland, Italy, Japan, Luxembourg, the Netherlands, New Zealand, Norway, Portugal, Spain, Sweden, Switzerland, Turkey, the United Kingdom and the United States.

* * *

TABLE OF CONTENTS

Foreword 4

Introduction 5

Part One

THE SOCIAL DIMENSION OF SCIENCE AND TECHNOLOGY

I. Society and Scientific Progress 22

II. Science and Social Change 34

III. The Social Function of Science and Technology 44

Part Two

SCIENCE POLICY AND
THE MANAGEMENT OF NATURAL RESOURCES

I. Natural Resources : A Complex System 65

II. Science, Technology and National Resources Policy . . . 75

III. International Relations 87

Part Three

TRENDS IN THE LEVEL AND PATTERN
OF R&D EXPENDITURES IN THE OECD AREA SINCE 1971... 99

Annex :

A. Graphs and Tables 108

B. Sources and Methods 119

CONCLUSION 131

FOREWORD

This volume contains the principal reports prepared for the Meeting of the OECD Committee for Scientific and Technological Policy at Ministerial Level, which took place in Paris on 24th and 25th June, 1975. The Introduction consists of the Report of the Committee submitted for discussion by Ministers. The issues presented in this Introduction were analysed in greater detail in two Background Reports, "The Social Dimension of Science and Technology" and "Science Policy and the Management of Natural Resources", which constitute Parts One and Two of the present volume. In addition, the arguments advanced in the Introduction were founded also on an analysis of the trends of research and development expenditures since 1971, the date of the preceding Meeting of Ministers of Science of the OECD Countries. This analysis is presented in Part Three. Finally, the conclusions arrived at by Ministers, as presented in the Communiqué issued at the close of the Meeting, are given in the Conclusion.

INTRODUCTION

The Preceding Meeting of Ministers of Science

The general theme of the Meeting of Ministers of Science of the OECD countries held in 1971 was "Science Policy in the Seventies". The report prepared by the Secretary-General's ad hoc group - Science, Growth and Society ("Brooks Report") - served as a reference document for Ministers' discussions and helped to define the problems which were emerging and to discern new trends and their implications for both society in general and for decision-makers, concerning, in particular, policies for science and technology but also other areas of governmental responsibility.

The discussions focused on two main considerations : the changes in social goals and the increasing preoccupation with qualitative aspects of growth called for a new orientation of science policies designed to correct, control or avoid the undesirable effects of rapid technological progress and to respond to collective needs. Second, it was realised that decisions relating to scientific and technological research could no longer be treated separately from other aspects of governmental decision-making, and that close liaison should exist between policies for science and technology and all socioeconomic concerns and governmental responsibilities.

This realisation of the need to change both the objectives and the means of science policy was already implicit within the context of a relative levelling-off of R&D expenditure in many Member countries. Re-orientation of research programmes and institutions appeared therefore all the more difficult.

It is now important, in mid-decade, to review these trends in the light of the major events which have transformed the social, economic and political scene. Since science and technology are of fundamental importance for the management of the present, as well as the construction of the future, policy-making in this field has a special role and responsibility in the process of understanding and mastering the new problems raised by the present situation and the decisions which must be taken in order to prepare for the future.

The Background to 1975

In 1971, all OECD governments were committed to a policy of economic growth, and it was expected that the growth in GNP during the decade would at least match that of the 1970s. Since the last Meeting events, and indeed crises, have clearly demonstrated that industrial societies have reached a stage of development characterised by the existence and emergence of interrelated difficulties, for which conventional policy instruments and institutional structures may no longer be adequate.

At the centre of the debate on the future of industrial societies is the constantly recurring question of the limits to economic growth. And the crisis triggered off by soaring oil prices has shown that, even if such limits still depended mainly on economic and political factors, it is no longer possible to disregard the contradiction between the exploitation of natural resources, environmental constraints, and economic growth requirements on the one hand, and the rate of increase in world population on the other.

Many of the symptoms of these difficulties were already discerned and discussed at the last Meeting of Ministers of Science. Subsequent events have not only confirmed the trends foreseen but have gone far beyond the degree of change expected : inflation has reached alarming proportions, and the threat of rising unemployment looms large over societies which had barely grown accustomed to full employment. It seems that expansion no longer holds out the reassuring prospect of a "game in which everyone can win" ; on the contrary, disparities within and between countries are becoming increasingly apparent.

One of the main conclusions of the "Brooks Report" is even more relevant today : in order to overcome these difficulties , further interventions in the operation of the market economy will become necessary whether in regard to industrial development, environmental concerns or collective aspirations. Whatever the trend of oil prices, the energy situation will provoke radical transformations in industrial and technological structures ; in the medium and long term substitution of certain materials and techniques will increase and must be controlled and directed so that they can be implemented with the minimum economic and social cost.

The circumstances under which the corresponding adaptation and re-orientation of research programmes and institutions are to be achieved are therefore even more complex than in 1971. New research programmes are not likely to be spectacular operations, but will require the integration of knowledge and techniques deriving from many branches of the natural engineering and social sciences. At the same time, public opinion now has greater reservations about science and technology and is more anxious to have a say in the use made of them. Science policies must take account of these new partners (consumer groups, public advocate groups, environmental groups, etc.) which, insofar as they perceive the possible consequences of technological development more acutely, wish to influence the process of decision-making.

The rapid pace of change and the appearance of "crisis" situations have led to increasing uncertainty, and even fear, in the face of the future. The multiplicity of models of saturation, if not the prophets of doom, are among the many symptoms of this anxiety. It is becoming increasingly apparent that present economic and political systems must be adapted to meet what have been referred to as the "internal limits" - i.e. the basic human needs of an expanding world population - without violating the "outer limits" of our planet's resources. Though the balance of world market forces has been altered as a result of the energy crisis, the disparities between rich and poor countries are increasing at the very moment when international relationships are characterised by their interdependency.

The basic problem facing governments is contained in the formula, "interdependence and complexity", which underlines both the possibilities and the limits of science policy. The "management of complexity" calls for a new outlook and approach to overall policy-making. This means that science policy is now operating within a very different political framework in which it has a special role and responsibility that extends beyond the utilisation of science and technology in the implementation of overall or sectoral policies. The challenge of relevance - how R&D activities can be better applied to the solution of the real problems of society - is inseparable from the political challenge of the definition of goals - how can the policy-making process take better account of these problems, as expressed by an increasing number of social partners.

Science and technology for the management of complex problems

Three basic requirements are necessary for science and technology to fulfil their role in today's situation. The first is understanding of the present structure of society and, particularly, of the respective roles of government and industry. The second is to define the new imperatives for government imposed by the evolution of society. The third is to assess the potential contribution of science and technology to the needs of society, government and industry in adapting to the perception of the future.

The boundaries of policy for science and technology are difficult to define and each of the two components, science and technology, raises specific considerations for overall policy-making. The growth of science and technology is accompanied by undesired and unanticipated results; hence equal importance must be given to the development of new technologies and to the means of remedying their possible negative consequences. This in no way implies that science and technology can provide the solution to all problems, much less that science policy is the key to overall policy. The fact remains, however, that science and technology pervade all aspects of society and are, moreover, one of society's main sources of innovation and dynamism.

Science and technology have a "social dimension" which calls for a new conception of, and approach to, the use of the scientific and technological potential. What is at issue is not only the management of science and technology, nor even their integration within the overall policy process, but the maintenance of our civilizations in the face of the disparities and conflicts within and between countries. It is not crises that should bring about the necessary changes in behaviour, but foresight, understanding and will. The real problems are not created by science and technology but by human behaviour, social systems and the relations between countries. But the challenge posed by any or all of these problems will not be met without the contribution of science, nor without deliberate government involvement in an open debate on alternative solutions.

This is why the Committee for Scientific and Technological Policy feels that the issues presented for discussion by Ministers cannot be related solely to scientific and technological factors, because they, in fact, intersect other areas of government responsibility, and underline the need for science policy to be regarded as an integral component of government policy. These issues are, therefore, addressed to Ministers in their dual capacity as responsible for all, or part, of the scientific and technological system and as participants in the formulation of overall policy. For practical purposes, they have been grouped under the six headings which follow, but, in fact, all relate to the fundamental problems confronting society - the management of complexity and the interdependence of the factors involved in all policy decisions.

I. THE ROLE OF SCIENCE AND TECHNOLOGY IN POLICY-MAKING

How can the potential of science and technology be effectively used in policy-making ? How can it be utilized in the definition of government objectives and in the implementation of decisions ? These questions have been posed ever since the initiation of national science policies : the continuing debate concerning the proper balance between a centralised and pluralistic model. Experience shows that no national system of science policy has followed an idealised form of either model, but instead, the pendulum has swung between the two approaches.

Whatever the solutions adopted, the difficulty is to determine how the bodies responsible for science and technology can not only define their own objectives, but contribute to the definition of overall government options. Many examples could be quoted which demonstrate that the close involvement of science policy bodies in the overall government decision-making process has not always been satisfactorily achieved. The questions raised are whether the inability of science policy to play an effective role is because the scientific and political communities do not have a common language ; "scientific" solutions proposed to policy-makers have failed to take into account the political repercussions of these solutions, or whether it is due to the fact that science policy machinery is too far removed from political decision-making centres.

Policy decisions, whether taken by the public or private sector, become increasingly irreversible because of the vast human and material resources involved and the number of people directly or indirectly affected by them. This leads, if not to outright resistance to the prospect of change, to the maintenance of the _status quo._ This points to the need for a means of determining the compatibility of options before they are adopted and of anticipating problems which may result. Long-term prospective planning might be one such tool. The multidisciplinary nature of such planning raises the question of the particular role of Ministers responsible for science and technology in formulating long-term government strategies.

Long-range prospective studies and plans must take account of the interests, motivation and requirements of the parties primarily concerned. In view of the growing interdependence of Member countries' economies, both among themselves and with the rest of the world, these long-term studies will need to integrate data describing the international environment. Some form of prospective analysis must, therefore, be undertaken at international level in order that information specifically needed for this type of research can more easily be exchanged and world trends analysed.

II. THE MANAGEMENT OF THE RESEARCH SYSTEM

Research and development in OECD countries is now passing through a difficult period. In terms of the available figures, it would appear that R&D expenditures at constant prices have been declining in a number of OECD Member countries. In fact, the percentage of total government budgets allocated to R&D appears to be decreasing in almost all cases, which may indicate a lower priority for R&D as a means of meeting government objectives.

Such aggregates, however, do not reflect the shifts in the structure and direction of the scientific and technological enterprise. For example, it seems that significant changes are taking place in the relative importance accorded to energy, defence and social R&D.

Increased allocations of resources to certain sectors of research will not by itself guarantee the production of relevant results and their timely application. The problems which arise are not always due to lack of financial resources but to the outmoded or inappropriate nature of decision-making structures. Moreover, when priority is suddenly accorded to one research sector or another - for instance of energy - the negative effects of a "stop-go" policy become apparent. In fact, both the constraints imposed due to lack of funds and the comparative affluence brought about by the urgency of new policy problems are the source of the same pitfalls. There is always the risk that policy-makers will be tempted, in one or the other case, to sacrifice long-term research, whose applications may appear uncertain or of doubtful relevance, to present exigencies. Research is still an imperfectly understood territory where ideas are generated and transformed. It cannot be subjected to rigid administrative rules or to very narrowly defined policy needs. Certainly, choices have to be made. Lack of knowledge of the relationship between science and society makes these choices more or less arbitrary, with the accompanying risk of limiting future options.

The fact that the measures taken in the past have been based on somewhat fragmentary knowledge makes a re-examination of the components of the system all the more necessary. Such a reappraisal could lead to the definition of new directions for research : it carries the risk, however, of introducing new factors of rigidity if those involved - actors and institutions - do not help to shape the new policies.

Scientists and research institutions have frequently fulminated against the obstacles which limit their autonomy and prevent them from being as innovative as they would wish. It would appear, however, that in many cases research institutions do not fully avail themselves of the autonomy - however limited - accorded to them. Yet the evolution of industrial societies and the increasing part which research is called upon to play, accelerates the pace of change and outstrips the speed of decision-making processes. The production and utilisation of new knowledge are in constant evolution and require rapid adaptations on the part of all concerned. Clearly, research policy must be founded on the dynamism and imagination of the individuals concerned, at all levels of conception and execution, if it is to be successful.

III. SCIENCE, TECHNOLOGY AND SOCIETY

In the exercise of its political and social responsibilities, science policy is faced with a series of fundamental problems such as the co-existence of spectacular technological achievements and long-standing, emerging or unsatisfied collective needs, the pervading impression of an inequitable distribution of the benefits and dis-benefits of technological progress, a technological development which broadens the range of options open to decision-makers but also increases the possibilities for error.

Obviously, science alone cannot solve conflicts of values and interests within society. The fact remains, however, that scientific knowledge has a short-term or long-term impact on most - if not all - social concerns. Over the long term, public perceptions, attitudes and values evolve as a result of technological progress, as well as conceptual changes and new awareness of social situations brought to light by research, as in the case of birth control. Over the short-term, all sectors of policy could benefit by calling upon science to assist in the formulation of goals, and in the implementation of programmes. Thus, government has a responsibility for ensuring that relevant knowledge is taken into account in the on-going social debate, and that it is organised and integrated to respond to the needs of decision-makers at all levels.

It has become as important to learn how better to utilise the growing stock of knowledge produced by all disciplines, as to increase that stock. It becomes, therefore, especially important to assess the potential relevance of the available scientific and technological information. The development of effective methods of utilisation, at the levels of collection, analysis and synthesis of data, is an essential pre-requisite for decision-making, taking into account scientific and technological advances. The reformulation of social or political concerns in terms amenable to research, the translation and integration of research results into policy conclusions or proposals have become, for example, specialised activities, leading to the development of professions or institutions which act as intermediaries between the parties concerned.

The association of scientists - including social scientists - in the formulation of policy objectives, as well as the increasing involvement of policy-makers in the determination of research orientations, also evidence the same trend and reflect the same urgent concerns. The social sciences and the natural sciences, however, co-exist at present in two intellectually separate worlds : there are, in fact , "two cultures", sometimes antagonistic, at least foreign to each other. Neither technological nor social problems can properly be tackled in all their complexity whilst this situation persists : when social goals are uncertain and changing, it is important that their formulation embody the most up-to-date available understanding of the interactions between technology, the economy, social trends and the quality of the environment. In attacking problems, disciplinary specialities must be brought together to allow for closer association of the natural sciences, the social sciences and the engineering sciences. It has sometimes proved difficult to maintain throughout the duration of a research project the same standards of relevance to social concerns that prompted its launching.

It has also proved difficult to identify and integrate the disparate knowledge already made available by the various disciplines. It should be recognised, however, that the perception and appreciation of the role of the social sciences by scientists from other disciplines is largely conditioned by the extent to which social scientists are able to carry out research leading to persuasive, significant and relevant results. This requires a reassessment of social research, in terms of both objectives and the resources made available to social scientists. A fuller integration of the natural and social sciences would seem, therefore, to call for research policy more carefully balanced to take the social sciences into account.

Public concern about the ethical, political, economic and social implications of the application of research being carried out in such fields as genetic engineering, weather modification or pharmacology, is increasing. In some cases, scientists themselves have been alarmed by the possible misapplications of their work and some have reoriented their activity towards other fields of research. In other cases, public administrations have been tempted to shy away from the support of certain types of research. This situation may lead to unexpected and undesirable underexploitation of certain fields of research, or to the unnoticed progress of research which may have harmful applications.

Too often, it is not until the importance of the resulting ethical political, economic and social problems becomes apparent that society is led to examine the implications of the development of a particular technology. In order to avoid such belated awareness of the implications of research, it is desirable that a "dialogue" should be established from the moment that the possible consequences of research results can be foreseen. Scientists themselves are in a position to issue early warnings to society about the implications of their work and should recognise their special responsibilities in this connection.

It is becoming increasingly urgent to develop not only methodologies and mechanisms for the assessment of technology, but to define the functions of research policy in the stimulation and control of technological progress. A true assessment of technological progress, coupled with the concern to mitigate its undesired effects, calls for the joint efforts of government, industry and the public. New needs are emerging to reconcile the aims of policies for the stimulation of technology with the need to control its undesirable social effects.

In addition to the broadening of the scope of technology assessment, its function also appears to be changing. Not only is the concept being applied to define whether socially acceptable conditions can be found for a government or industrial initiative, but it is being expanded to determine which initiatives should be taken. The long lead-time required for application, the relative inability of governments to stimulate technological advance in one or another sector - for example, industry, the administration, public services - would seem to indicate that certain social needs are poorly or insufficiently taken into consideration by research policy. Conversely, research policy does not appear to be equipped to influence certain types of economic, social and political use of knowledge.

IV. THE NEED FOR PUBLIC INVOLVEMENT

The pace and character of technological change have given rise to increasingly outspoken public misgivings. Although many resulting benefits are recognized - in terms of better living conditions, better health, greater expectation of life, more general affluence and comfort - perception of the disparities and risks which can accompany such change is becoming more and more explicit.

Disparities result from the uneven distribution of the results of science and technology throughout society. The promotion of innovation and experimental work directed to social objectives seems to lag behind the technological accomplishments geared to strategic or industrial ends - at a time when all societies are increasingly vulnerable to unexpected shifts in the international political equilibrium or to internal disruptions generated by localised social protest.

The negative aspects of technological change have been underlined in recent years, as events have confirmed previous predictions and given greater credibility to current projections. Damage to the environment, with potential harmful consequences for the health and well-being of human life, has multiplied. The uncontrolled development of industrial technology has contributed to dissatisfaction with the conditions of urban and working life.

Decisions have been made and actions taken to cope with these difficulties, and some positive results are already apparent. On the other hand, however, decisions are also being made or considered which seem to inspire new fears among the public that governments are more likely to respond to crisis situations than to anticipate and prevent their occurrence.

This more general perception of the risks inherent in technological change is to some extent a tribute to scientific research. Progress in the natural and bio-medical sciences has led to greater awareness of the physical consequences of the introduction of new technologies. Progress in the social sciences has led to more intimate knowledge of society - with the result that social disruptions are more accurately and more rapidly identified.

This increased perception of possible negative consequences is not without its own dangers : any political action implies a certain element of risk. Too sharp a perception of the dangers inherent may obscure possible benefits, and lead to the inability to decide and to act. Governments, who are normally more sensitive to immediate benefits, are in the position of making decisions which will be scrutinised by a public which is now becoming more responsive to long-term and uncertain dangers. There is now, therefore, the making of a "stalemate society" unable to act because it cannot reach a consensus on what to do and how to harmonize national strategic goals and individual aspirations.

This stalemate can be overcome if the application of scientific knowledge and the direction of technology become the subjects of a continuing debate between national and local governments, the scientific community and the public. Thus, technological problems may stimulate the desire for democratic participation in decision-making which has become a central subject for political debate and social demand in many countries.

There is another dimension to this development, relating to the very structure of technological change. There is strong evidence that pluralism in the discussion of scientific and technological strategies is, in itself, a factor of progress. Such pluralism permits a more systematic identification of social problems, contributes to a sharper definition of alternatives and allows for continuous monitoring of socio-economic change.

Social acceptance of technologies has become an essential prerequisite for their successful implementation, as evidenced by public reaction with regard to nuclear energy. Public involvement in governmental decision-making and better public understanding of the contraints under which governments must operate are thus necessary conditions for the future balance of national scientific and technological efforts. Scientists and technologists must find their own place in this dialogue between government and the public. The introduction of new technologies more and more often calls for review of conflicting assessments of their possible benefits and socio-economic costs. This makes clear the need for scientific and technological expertise, sufficiently free to express views independently of the various interests involved.

The demand for public involvement in decision-making assigns another function to science - that of exploring the means by which such participation can be realised. The difficulties involved in developing such participation have become increasingly apparent, and this question should therefore be the subject of extensive social-science research.

V. A NEW VISION OF NATURAL RESOURCES

For some years past, a series of difficult situations and crises has clearly demonstrated the fundamental importance of natural resources, not only from the point of view of the economy and of the quality of life, but also for the survival of the industrialised as well as the developing countries. The energy crisis is today the most visible problem because of the political and economic aspects of this problem, but it should be seen in conjunction with the abrupt fluctuations in prices of certain raw materials, the recognition of the dangers inherent in the deterioration of the environment and the shortage of food resources with which almost half of humanity is confronted.

These difficulties and crises arise because growing anxiety about the exhaustion of the planet's resources, coupled with expotential population growth, aggravated by political and economic factors, introduce a new element of geo-politics into questions of access to, and use of, these diminishing resources. The better management of these resources is therefore of vital importance.

Science and technology have no ready-made solutions to offer in the very short term. But this does not mean that they have not a very important role to play : examination of the potentialities of research and development shows quite clearly that they can offer many possibilities in the medium and long terms. However, these possibilities cannot be realised unless a large scale R&D effort is planned and put into effect now, on the basis of a new vision of natural resources.

Within this new perspective, natural resources should be regarded as an endowment which is limited and which should be managed in such a way as to satisfy the needs of all humanity and of generations to come. The second point is that all natural resources (energy, materials, food, water, air, land) are closely interrelated and therefore form a complex system with many dimensions (technological, economic, social, political). Moreover, the interdependence of countries is also a basic feature of the system of resources. There exists therefore a clear need for an overall resources policy.

Science and technology can assist in the formulation of such policy by better defining the nature of the problems and considering all the interactions involved. This requires greater understanding of the complex system of resources in order to be able to identify the choices available and to assess the implications of these choices. Knowledge of the system constitutes, therefore, in itself, a priority research objective. In addition, the time horizon and the applications of any R&D activity or industrial application are such that decisions must be based on possible medium and-long-term developments in the field. Anticipatory or "strategic" planning is therefore particularly essential. Responsibility for such planning rests primarily with governments.

The field of natural resources embraces many aspects and problems which are international in character. This international dimension must be taken into account by governments in promoting the research aimed at a better understanding of the system, and in the "strategic planning" mentioned above. There is, therefore, a need for co-ordination of national efforts related to these two tasks.

Science and technology have also to contribute to the implementation of an overall resources policy by keeping open a sufficient number of alternative options, as well as making it possible to achieve particular objectives. In view of the broad extent of the resources field and the multiplicity of the research directions, choices must be made and priorities established, taking into account their implications for the overall system and the time-frame within which the corresponding efforts could be achieved.

In this respect, an important problem is due to the fact that, in the public sector, responsibilities for different aspects of natural resources are dispersed among numerous government departments with different and sometimes conflicting functions (e.g. environment and industrial development). This situation can lead to inconsistencies, even contradictions, in governments' science, economic and social policies.

In the field of natural resources, industry plays a predominant role and carries out most of the relevant research and development. This raises the problem of the harmonisation of the objectives of industry and those of an overall natural resources policy, when market mechanisms are insufficient for this purpose. This problem is particularly important as regards the support of R&D on those alternative options which are not necessarily the most economically advantageous at a particular moment in time. Another important example is the question of resource conservation, which is a critical factor in all sectors of the natural resources field but does not necessarily constitute a "natural" R&D goal for industry. The same considerations also apply to the protection of the environment.

The contribution of science and technology to a policy for resources is, of course, dependent on the amount and quality of scientific and technical personnel (scientists, engineers and technicians) in a wide range of disciplines. It is therefore essential to ensure that future needs for such personnel are realistically assessed and adequately catered for by higher education institutions.

VI. METHODS FOR INTERNATIONAL CO-OPERATION

As a result of the energy crisis, the interdependence of all countries has become abundantly clear, particularly as far as natural resources are concerned. It has also led to a profound evolution of international relations, and their political, economic, financial and monetary aspects. Within this framework, science and technology are called upon to play an increasingly important role because of the possibilities they can provide in the medium and long term both for dealing with the crisis and for the better management of resources in order to satisfy the needs of all countries. It is therefore now opportune to re-examine the question of scientific and technological relations.

Whilst the amount and nature of their indigenous resources vary greatly, a common feature of the OECD countries is their highly developed scientific, technological and industrial potential. Even though their short-term economic and political interests differ greatly, many factors point to the need for increased co-operation in science and technology between Member countries, particularly in view of the fact that, in the case of natural resources, the increased R&D effort which will be necessary goes beyond the possibilities of individual countries.

Intergovernmental relations between OECD countries with respect to scientific research are already well established, but such is not the case for technology which should therefore be the subject of particular attention, taking fully into account the technical and capital risks involved, as well as the implications for national economies in the context of increasingly competitive conditions.

The development of international co-operation in technology between all the OECD countries within the framework of a common overall policy for natural resources would be extremely difficult to realise at the present time because of the great differences in their situations with regard to natural resources. It could, however, be envisaged on a more restricted regional basis. A more pragmatic approach is, of course, co-operation on an "à la carte" basis, such as exists at present, where the initiative originates from individual firms, resulting from their particular industrial and commercial interests. This approach has, however, limitations in that it does not correspond to the need for overall policies, where governments have a central role. Consideration might therefore be given to whether, and how, some international division of technological research and developments might be arrived at, taking into particular account the complex question of the sharing of risks and economic benefits between the parties concerned.

Whatever the form of international co-operation envisaged, account must be taken of the fact that industry is the principal performer of technology related to natural resources. This raises the question of the respective roles and relationship of private enterprise and government in the international context (particularly with regard to the role of multinational corporations), since governments will be called upon to provide investments and such measures as market guarantees that, in the aggregate, are beyond the possibilities of individual countries.

Science and technology are, in fact, a resource exclusive to the industrialized countries, and therefore represent an increasingly valuable asset in their relations with the developing countries since the scientific and technological skills of the former are necessary for the economic and social development of the latter. A new pattern of international relations can therefore be seen to be developing, based on complementary interests, in which the OECD countries can provide scientific expertise and technologies in exchange for security of supply of raw materials.

On this basis, exchanges mutually beneficial to both parties can be developed, but it must be borne in mind that technologies have, for the most part, been developed to meet the needs and conditions of the industrialized countries. This raises the important question of the choice of the technologies most appropriate for the developing countries, and of their adaptation to local circumstances. It must also be recognised that the transfer of technology will not, by itself, lead to a process of industrialization and economic development, for which a whole set of conditions are necessary (in particular, the training of qualified manpower), often requiring a long period of time for their fulfilment. Hence, the need for a continuing dialogue between the industrialized and developing countries which takes account of these factors and their medium-and long-term implications.

Part One

THE SOCIAL DIMENSION OF SCIENCE AND TECHNOLOGY

> "Science appears as what in truth she is,
> Not as our glory and our absolute boast,
> But as a succedaneum, and as a prop
> To our infirmity."
>
> W. Wordsworth (1770-1850)
> "The Prelude".

To affirm the social character of science and technology is to state an obvious truth. Scientists and engineers work within the social body ; governments and firms finance their activities in relation to needs perceived at various levels of society ; and it is in the social crucible of interests and pressures that results of science and technology are, or are not used.

Defined in this way, the social dimension of science and technology is not a recent discovery. What does appear to be a new development, however, is the reappraisal of the relations of science with society, which is expressed in terms of the political will to define, influence or modify those relations.

Doctrinarians of Scientism in the last century had regarded science and technology as a decisive factor for human progress. The combination of society's concerns and the logic of scientific development was, through successive stages, supposed to result in a better social order. Leading to the mastery of social relations and of society's relations with nature, increasingly sophisticated technology was supposed to ensure continued progress.

The Management of Natural Resources illustrates the contemporary questioning of a certain view of progress, through growing awareness of the natural limits which have to be taken into account (1). To these natural limits must be added the increasingly felt social limits : the change brought about by scientific and technological development is not necessarily a factor of "progress" when it is responsible for setting off a chain of dislocations within societies already convulsed by the growing pace of technological development.

The future of industrial societies is more and more closely geared to the evolution of knowledge. The social function of research was for a long time too widely dispersed to be truly perceived as such. It is

---

1) Cf. Part Two.

nevertheless ever more clearly expressed as a threefold attempt to manipulate nature, mankind and society. In these circumstances, it is hardly surprising that knowledge itself should appear to be a last resort against the problems which its own uncontrolled exploitation has caused to appear, or has at least rendered more acute. Science is no longer the good fairy hiding in the wings of the economic, social and political stage, who distributes her goodies by an almost automatic process which it has sometimes seemed possible to accelerate but not to modify. Science now holds the front of the stage and plays one of the leading roles in an as yet unfinished drama. Science, which is the source of both hopes and fears, can increasingly be seen as the reflector, or the detector, of the major problems of society.

Therefore, science has become open to social conflict and is now one of the stakes involved. Already, everything we owe to it is fading in the memories of men, who now think only of the perils and difficulties of the present. The accumulation of social problems leads to much being expected from the combined contribution of the natural, engineering and social sciences. The results of the efforts made in this field are disappointing, however. The solutions found through technological research carried out over the last few decades are not always suitable for new situations. Does not giving new impetus to science policies and directing them towards new goals call for a review of the postulates of such policies ? Should not the structures of science and of the knowledge resulting from it be re-assessed in a new social context ? Is it not necessary to measure more accurately the social consequences of the science and technology effort before proposing any remedies ?

I

SOCIETY AND SCIENTIFIC PROGRESS

Many difficulties confronting industrial societies would appear to be due to uneven development which has left its marks on them in the past and continues to do so even now, to the point that inadequate adaptations in some areas is blocking or even paralysing the evolution of the whole social body. Thus, the political history of modern States has given birth to bureaucratic institutions which are ill-suited to new circumstances and often constitute barriers to change. At the same time, society is imbued with mental and moral attitudes which survive well beyond the social conditions responsible for their first appearance.

The structures of science itself may seem to be falling behind in relation to the expectations of society.

It is easy - but unfair - to put the scientists themselves on trial for this. Scientists and engineers work in institutions and according to practices which reflect a compromise between their professional demands and the aims of society. The relations of science with the community were marked, immediately after the Second World War, by an optimism which dissimulated many of the difficulties that have become more apparent today : it might have been thought that because they were concerned with science and with research, scientists would be more innovative than others, and would be more capable of modifying their institutional framework, according to their needs. This was to assume that they were wiser than the members of other professions. It also credited them with more ability than they possessed to influence events : experience has shown that a handful of scientists is not enough, for example, to revive an academy, change a university, or influence a firm's policy.

On the contrary, industrial society's debt to the scientists and engineers who have served it so well in spite of the imperfect structures must be recognised. It is in the domain of ideas, of the formulation of research topics and of the exploitation of results that the dynamic spirit and imagination of contemporary science and technology have acquired most prestige. It is not the quality of the men that is in question, but the defects of a system which makes success a question of chance, stifles social demand, does little to encourage collective efforts and prevents society from both systematically taking advantage of knowledge and forearming itself against its dangers.

A. FROM BACKROOM TO "SCIENCE COMPLEX"

It is the same with science and technology as with any other activity : usually when activity slows down or falls off, when laboratories are beginning to be closed and when recruitment stops, questions surface, which were previously latent or confined to a minority, as to the foundations of earlier prosperity. In particular, the phenomenon of levelling-off research and development expenditures leads to seeking greater efficiency in the use of resources. This in turn calls into question the ability of the scientific and technological system - as historically organised in relation to social structures and values - to provide a suitable framework for development which would enable it to make a real contribution towards achieving society's aims.

If such problems arise, it is largely because the fertility of scientific effort has exceeded all hopes : the increase in knowledge, the growing number of research fields and the appearance of new disciplines have made science an ever more complex enterprise. The limitation of resources means that they must be used with economy and demands more severe selection. But how is such a rich universe to be organised ? How is its activity to be oriented without impairing its dynamic force ? Such questions lead us to consider the scientific community itself and more particularly the scientists in the various disciplines who explore the frontiers of knowledge. Doubtless, such scientists are relatively few compared with the total body of scientists and engineers. But this community is of decisive importance in the system of knowledge production. It is increasingly aware of the major social concerns of the moment : in a period of shortage, economy of resources calls for more efficient forms of organisation. Society's difficulty in equipping itself to face its problems rebounds on the choice of subjects to research.

Just like those regiments which glitter for inspection purposes or those government departments which put on a new façade for one day in order to receive a minister, the scientific communities in industrial societies have an appearance which bears little relationship to reality.

The appearance : cohorts of scientists and engineers in all disciplines and with all kinds of institutional affiliations surmount the wide diversity of their interests in order to throw back the frontiers of knowledge and mobilise it on behalf of society as a whole.

The reality : a change in outlook and in the ways in which scientific work is organised is already perceptible but it still allows disciplines to persist which, clustered around their own intellectual or professional interests, do not always maintain close mutual relations and sometimes only invoke current political, social or economic objectives as a pretext to pursue existing research programmes. This attitude seems perfectly legitimate, in any case, to those who think that "what is good for science is good for society".

The spectacular increase in the scientific and technological activities of industrialised countries has profoundly changed the conditions of scientific work, although scientific communities have not completely adjusted to this new state of affairs. A short while ago, it was thought that industrial research constituted a model in this respect which other sectors could take as an example ; the disenchantment felt by many firms for their own laboratories only serves to increase the uncertainty. The feeling that there has been an uneven adjustment of the research system combines with the confusion resulting from criticism of the "models".

The attitudes of scientists and the analyses made by the sociologists of science usually coincide, however, in continuing to place scientific life in the context of the past : as though the efforts which involve powerful agencies and hundreds of thousands of men could be reduced to those small élite communities in which a few scientists in each discipline could easily follow the international progress of knowledge in their field and mobilise their colleagues, in exceptional circumstances, to serve society. Scientists are of course still able to follow developments in their fields of interest - but the scientific territories of the various specialists adapt to new circumstances by shrinking like the piece of shagreen leather in Balzac's story. There is a need to create once more élites capable of relating several fields of knowledge to each other.

The astonishing rise in the number of scientists during the 20th century does not therefore, by itself, reveal the extent of the changes that have taken place. Not only have scientific communities increased in number, they have become extremely varied.

As regards science itself, new disciplines or subdisciplines have emerged or are struggling to emerge. Within the most traditional fields, new equipment has been responsible for the development of particular skills or specialisations. Technology has asserted itself as one of the essential instruments of science : the scientist is disarmed without the help of the technician who provides him with access to a reactor or a phytotron, to such a point that it is increasingly difficult to distinguish the scientist technician from the technician scientist, and fundamental research from applied research.

The institutional bases of scientific life have become more manifold and more varied. The university is no longer the exclusive, jealous spearhead of scientific progress. Industry carries out its own advanced research which is sometimes unequalled. Local, national or international administrations have high-quality laboratories.

Social demand has caused new activities to hatch out around research in the strict sense. Society is no longer always expected to absorb spontaneously into its activities the knowledge acquired by science. Application, which was once the result of chance, people and circumstances, is being increasingly replaced by a utilisation policy which endeavours to take systematic advantage of the results of research : industrial consultants, marketing services and agents handling administrative formalities are increasing in number and becoming the natural extension of the research effort, to such a point that these agencies appear more and more frequently in the career profiles of research workers.

The size of the political, economic and social stakes involved has also profoundly changed scientific activity both upstream and downstream. Organisations have made their appearance to stimulate research and to try and orient it. Can any of them be arbitrarily isolated from the scientific community, understood as the community of research workers only ? The institutions which finance oriented or non-oriented research are run by scientists who are often bound to their colleagues by personal and professional ties, since they often come from research laboratories or think of returning to them some day, and since they assume their administrative responsibilities with the help of committees of experts which enjoy the authority attached to famous names.

Today, therefore, after the evolution of the last few decades, research is seen to be diversified, fragmented, extended and bureaucratised : it has indeed become a "complex". Teams and institutions in the full throes of renovation rub shoulders with others whose habits stay the same. In spite of the clamour from the outside world, many people in the backroom of research try to survive as though nothing was afoot.

New structures have of course sprung up which offer new career opportunities, but these in fact serve as safety valves and enable individuals who so desire to come out of the backroom, thus protecting it from internal pressures which would make it change. For the last two centuries, the system of rewards in this backroom has remained appreciably the same : scientific merit, judged by peers in each field of research, makes reputations and distributes prestige. Only a few individuals can take the front of the stage, in spite of the large numbers whose role is to justify and back up their work. On an international scale, a few scientists form "invisible colleges" in each field : but for many young scientists who do not succeed in breaking through, these colleges have the appearance of an insidious form of intellectual and professional exploitation.

Is research not therefore a ship without a captain, tossed about at the whim of science's "internal logic", which in the last analysis is usually no more than the reflection of a passing fashion or infatuation ? Given the mass of publications and the fragmentation of work induced by a system where everyone is trying to make a place for himself in order to enter in his turn into the "invisible college", the difficulties of teamwork and of interdisciplinary research are no cause for surprise. The best chance of success for a professional research scientist still lies in individual work.

A professional social structure determines to a large extent the course of scientific life : this structure is inherited from a stage earlier than the development of a scientific and technical life, and its suitability for present conditions is increasingly doubtful. An indication of this may be found in particular in the tendency towards the "fundamentalisation" of all applied research institutions, even in industry. This tendency of research work to lean towards fundamental research is not of necessity always a bad thing, but the least one can say of it is that it is not what is intended. The causes of this are many, but one of the main ones is no doubt the "nobleness" attributed to fundamental research as defined by the university system. The connection between teaching and research is of course always essential : it enables the teachers to keep in touch with scientific development, to prepare students' awakening to a vocation as research workers, to introduce them to the problems of research, whose findings they will be using in their professional lives, and, lastly, to provide the community with a vast potential of scientific skills. However, the connection between teaching and research cannot be regarded as a dogma, enabling a system to be perpetuated which is indifferent to the practical needs of society.

In several industrialised countries, the slowing down of scientific effort has increased the anxiety. Formerly, relatively substantial annual increases enabled research to develop in new directions with new men. A period of shortage and stabilisation might have proved salutary in allowing flexible structures the better to assimilate the changes occurring and to achieve renewal in depth. But it seems that the internal structure of the research system is not sufficient to guarantee the continuous renewal of research topics and teams. By consolidating the position of the "mandarins", financial difficulties favour individual work at the cost of collective effort. Because of the faster progress of knowledge, the nineteenth century had been obliged to fractionate learning and to acknowledge the authority of specialities, becoming aware of the interactions between them and beginning to feel that specialised fields of knowledge can no longer be adequately mastered save by groups and teams which alone can keep up with the scientific and technological "state of the art".

Impatient with the slowness of the scientific world in adapting to the attempts to organise or orient research, policy makers often resort to authoritarian or bureaucratic procedures. Industrialists, for example, shut down advanced research laboratories without allowing them time to bear fruit ; government authorities are inclined to resort systematically to short-term contracts or to forms of financing which demand the formation of research teams. In all these cases, measures are taken without their consequences being really assessed for lack of sufficient knowledge of the mechanisms involved. Contractual procedures, in particular, have never been systematically evaluated from the point of view of their scientific and institutional effects. The proliferation of such contracts, leading to the recruitment of research staff for short periods, may create a category of "sub-researchers" and may be more appropriate for scientific work in one particular discipline than in another. Moreover, the extension to the social sciences of the contractual procedures applied in the case of the natural sciences and engineering probably does not take into account the circumstances peculiar to social research.

Thus, in the last analysis, the scientific world's resistance to the social pressures brought to bear on it also has positive aspects : for instance, a bureaucratic attempt to reorganise research might compromise the long-term equilibrium of the scientific effort if it really had the desired effects ; and, for want of something better, the permanence of the traditional standards protects the quality of scientific work. But the closing down of a laboratory is often merely an appearance, allowing researchers to continue their work under another label ; contracts are often no more than disguised subsidies and contract-chasing enables laboratories to stay afloat despite the winds and tides of the passing economic situation. But what useless and scientifically costly efforts these are, for survival in such a system !

B. KNOWLEDGE IN FRAGMENTS

In some countries, science policy is regarded as counting all aspects of scientific knowledge, including the social sciences and even in some cases the "humanities" : for instance, a professor of Latin or Greek has been known to preside over a government council on science policy. In others, science policies are concerned only with the natural sciences and technology. Some attempt has recently been made to extend them to the social sciences, especially where these can help to define and implement science and technology policies.

These differing concepts are not without their effects on scientific life itself. Being dominated by representatives of the natural sciences or engineering, science policies tend to relegate the social sciences to the field of cultural policy or, when they are concerned with these disciplines, to favour the empirical approach and quantitative analyses. Such policies tend therefore to impose on the social sciences a restrictive concept of science which is borrowed from the reasoning processes of certain natural sciences.

Taking the opposite approach, other policies tend to institutionalise jointly the social sciences and humanities. The former then borrow the literary approach of the latter at the expense of field research.

In spite of these differences, which are due more to the organisation of scientific research than to science itself, there is no apparent break in the continuity of knowledge and it covers a continuum ranging from physics to anthropology which has no clear-cut frontiers. Such frontiers only take on substance at the level of research organised in the light of their existence, to the point of resisting any attempt at grouping.

Interdisciplinary research - where the work of scientists from different disciplines really merges - is still an always hoped-for but rarely achieved ideal, while multidisciplinary approaches themselves - which allow research workers in different disciplines to explore the same problem side by side and to compare their results - are still the exception. These difficulties encountered in carrying out research only emphasize the fragmentary appearance of the results obtained. How can the product of medical research be integrated with the results produced by sociological or psychological research ? How is the knowledge obtained by the social sciences to be used when evolving new technologies ? How, for instance, before undertaking further research, can a full account be obtained of the state of the art where the problems of energy production, conservation and consumption are concerned ?

It is therefore as much a matter of making better use of the stock of knowledge as of increasing it. The available technical knowledge is not always used. Moreover, the future of society may depend as much on advances in our understanding of human beings as on advances in the natural sciences : the establishment of new relations between man and his environment is not merely a technical problem, but calls for new forms of individual and collective behaviour, deriving from changes in ethical standards and a better understanding of the social phenomena involved. It is not that there is too much knowledge, but that knowledge does not advance sufficiently on every front and is not always easy of access.

A joint contribution by the natural sciences, engineering science and the social sciences is increasingly essential to the study of any social problem. It is to be doubted whether the fragmentation of knowledge is a passing phenomenon which could be overcome by an effort of synthesis, and one may surmise that this fragmentation is a reflection of the heterogeneous nature of modern knowledge, which covers situations and phenomena that are so different - and so fluctuating where society is concerned - that to construct a general theory is now impossible.

The fragmentation of knowledge puts a brake on scientific progress by limiting disciplines' possibility of mutual enrichment. In turning its attention to social problems, research thus facilitates contact and interpenetration between different fields of knowledge. By submitting scientific knowledge to the test of society's practical problems, research orientation therefore facilitates a certain integration. But this process is all the more laborious and uncertain as social orientations are often not very explicit, and indicate targets which are themselves fragmentary and vague.

The only hope of overcoming such difficulties is at present offered by advances in analytical techniques which permit problems to be posed in all their complexity. Substantial progress has no doubt been made along this path. Mass education and modern information media have been responsible for some dissemination of fundamental concepts and scientific methods. Language - which enables social problems to be expressed - has been affected, and facilitates the dialogue between society and science. The value attached to the scientific method stimulates the development of procedures or analytical techniques - such as PPBS or systems analysis - which enable knowledge to be organized in the light of social concerns.

In point of fact, the most difficult thing for the consumer of knowledge, whether he is himself a scientist or the user of research, is not so much to obtain the information desired as to decide which information is desirable. When a problem arises, determination of the degree of pertinence expected from the information to be collected is, to a certain extent, necessarily arbitrary and the intuition of the person concerned - or, on the contrary, his ignorance - plays an important role. Attempts to present a framework for analysing problems are therefore particularly useful from this standpoint, especially as industrial society's craving for information finally leads to such an accumulation of observations that any knowledge of a really descriptive or explanatory nature is buried beneath them. Scientists are not the ones to blame : social demand, as expressed or deformed by institutions, is a mould into which knowledge is poured - or in which it is lost : the inability of institutions to look beyond their own concerns and consider the multidimensional nature of the problems facing them rebounds on to the structure of knowledge and contributes towards its growing fragmentation. In the last analysis, the fragmentation of knowledge will not be overcome by improved data processing techniques, but by a more coherent, accurate and complete expression of what society expects of it. Priorities and collective targets must be made more explicit.

C. SCIENCE AND ITS EXTENSIONS

The progress of our societies is indissolubly linked with science and technology. The products of research have yielded major improvements in the physical, economic and social well-being of the individual. Improved productivity has been reflected in industrialised societies in unprecedented material comfort and in a considerable broadening of the horizons of individual lives. Science promises man a future increasingly free from sickness, arduous work and ignorance.

However, today as in the past, the benefits are not equally distributed. Scientists are not responsible for this since they have often themselves been the victims of social systems which endeavoured to hold back the changes called for by their work. Every epoch has experienced a considerable gap between the well-being which was possible and that which was achieved, a contrast between the benefits and the evils of progress. But the present period is marked by the fact that economic and social development stresses some of these divergences, and scientific progress makes the consequences of technological change and the interdependence of society's problems more apparent. Human activity is on a new scale where its effects can no longer be disregarded.

The production of food provides an illustration. The distribution of adequate supplies - at least in the industrialised countries - was universally regarded as beneficial and having no harmful consequences. So long as no attention was paid to natural equilibrium, the utilisation of synthetic fertilisers, pesticides, hybrid species, machinery and large-scale irrigation projects was seen as having little adverse effect on the natural environment, the quality of agricultural products or the life of the agricultural population. It was the generous environmental reserve that existed, in combination with a certain set of human and social values, which supported the view that many areas of the human and natural environment were largely independent of each other.

It is now becoming increasingly apparent that the widespread and intensive development of technology interacts, often in undesirable ways, with other sectors of the environment. The resulting condition is characterised by a reduced ability to digest, absorb and adapt to the impact of technological change. This situation has led to a steady growth of interdependent elements, where both desired and undesired technological impacts exist side by side. There is a great deal of evidence for this, ranging from the expansion of the development of weapons to the breakdown of natural and social mechanisms as a result of modern industrial activity, and the economic vulnerability resulting from the importance of certain energy sources to industrial society. It is thus that the once open natural frontiers have recently begun to close, just as the geographical frontiers have done.

As interdependency grows, it becomes increasingly difficult to determine social costs and benefits. On this basis, the exploitation of science and technology brings with it the possibility of producing major damage. Yet most efforts to deal with the impacts - with certain exceptions - are of an after-the-fact nature. It is the uncertainty which science and technology have thus caused to be generated that has led to growing public pressure for the examination of technologies and their affects prior to their use.

Public concern towards the feared and undesirable consequences of technology, and the need to anticipate their effects, has now been extended to science. This results from the growing awareness of the relationship between science and technology and the products which derive from them. Large numbers of people - through education and communications - have some appreciation of the scientific underpinning of such products as jet aircraft, antibiotics and tranquillizers, synthetic materials and computers. Because these products are so important to modern society, the fundamental research which made them possible is increasingly regarded by public opinion as liable to hold in store other disturbing benefits. The effect of this kind of thought has created doubts as to the adequacy of ex-post-facto technology assessment as a means for coping successfully with unintended consequences. For, with the increased dependence of modern technology on science, still another and more fundamental precursor of human technical activities has been identified. This awareness has caused the transfer of concern from assessing existing technology to that which may be created. Anticipatory technology assessment then - insofar as technology comes from science - becomes almost equivalent to ex-post-facto science assessment.

The role of science as an agent of social change is still largely viewed through its contributions to the technology sector. These are made by exploiting research and through the training of manpower.

The reason for this particular utilitarian emphasis is due to the research-technology-product chain. Until recently, this chain was the most visible example of the link between science and its consequences for society. The natural sciences bear abundant witness to this. Fundamental scientific research has produced knowledge - in fact, actual formulae - which has led directly to products and technical processes, and has altered attitudes and influenced major policy decisions.

Several decades ago already, Pavlov's research had opened the way for techniques for manipulating the masses. The development of new synthetic materials threatens constantly to bring about profound economic upheavals. Organ transplants call in question accepted principles as regards the removal or organs, increase the cost of medical attention and call for a redefinition of death. Therapeutic tests on man are proving more and more essential for measuring the impact of certain products. Other products can be used to influence directly and predictably the behaviour of man. Lastly, increasingly accurate knowledge of the genetic code has probably brought us very near to the time when techniques will be perfected for manipulating it.

All these examples emphasise the extent to which scientific development has, in solving certain problems, been responsible for creating others which society is not equipped to tackle. But the gravity of these problems calls for a new approach : solutions cannot in fact be expected to spring forth spontaneously when technology wants them. Such solutions can only be found and accepted if they are the fruit of rational discussion, which must not be conducted under emergency conditions - when it is too late for the event. In view of this, some alarm system must operate as early as the research stage, when the possible results begin to appear.

Scientists and graduate engineers are the first who can draw the community's attention to the implications of their work. Social or scientific responsibility is very broadly contrasted here with the traditional concept of scientific responsibility : to warn society at the preliminary research stage is in fact to overreach strict scientific objectivity and to interpret the early results in the light of assumptions that may be fragile. Is there not a risk here of being pointlessly alarmist when a mistake may have been made ? Silence, however, would be more culpable than error ; the scientist is drawing society's attention only to a possibility. It is not for him to take a decision, which is always fundamentally a political decision. Already, various scientific organisations have examined these problems or have been set up for that purpose. There can be no denying that a new code of scientific ethics, which would appear to be in the making at present, would greatly facilitate the public debate needed before society is ready to solve the problems set for it by science. It is equally undeniable that steps taken by scientists could make it unnecessary for the public authorities to take measures of direct control over research which could be dangerous for its equilibrium.

These preoccupations are all the more timely as the research system has entered a period of profound adaptation, which multiplies the promises but also the risks of scientific and technological development, since this latter is now destined to increase its interaction with the public at large and with the political, economic and social institutions.

II

SCIENCE AND SOCIAL CHANGE

Scientific and technological development, as such, neither resolves nor creates economic, social or political problems, but it can attenuate or exaggerate them, making them either more, or less, tenable. Technology has exercised a profound equalising influence within our societies but in doing so it has sometimes revealed new aspects of inequality. Technology is not the cause of conflict, but the development of nuclear weapons has given the problem of war and peace a new dimension. Environmental pollution begins with human activity, but it is technology thoughtlessly put to the exclusive service of economic development that makes it a real menace. Genetic manipulation is only a danger in a society where abuses are possible. Medical care which is only available to an economic élite does not create such élites. In this sense, each society probably has the scientific and technological development it deserves.

By marking the solution of problems which hitherto did not threaten the future, more and more urgent, science and technology are the detectors of the disequilibria in industrial society. The existence of unused knowledge underlines the difference between what is possible and what is actually achieved and thus points to the limits of power. The improvement of living conditions reveals inequalities of a new type. Life has been made easier by the diversification of economic and social functions but society as a whole has been rendered more fragile.

A. THE INTERNAL CHALLENGE

Science and technology have multiplied society's means of action and reduced inequalities and factors of instability. Industrial societies thought to free themselves in this way from subjection to natural and social mechanisms. But it is becoming increasingly apparent that science alone cannot ensure social progress - though it can and must contribute to it and is in fact led by it rather than being its leader.

Scientific and technological progress well served those who knew how to profit from it. Just as the exploration of the New World enriched those who ventured to undertake it, exploitation of the new frontiers of knowledge is a source of prosperity and power. The conditions for success in this field are still little known : a mixture of the spirit of enterprise and managerial ability, capacity to change in response to the ways in which situations or problems develop, simple and well-defined targets whose achievement can be measured with the help of reliable and manageable indicators.

The ability to take advantage of the methods or findings of scientific or technological research is unevenly distributed. Only a very small number of countries in the world have been able to build up and exploit a scientific and technological potential. Among these few, moreover, some have proved more skilful in bringing their research efforts to fruition. The integration of science and scientists, of technology and engineers, with socio-economic activities can only be made fruitful by removing obstacles to the spread of ideas and innovations. These prerequisites are not equally met in every society and even, within each society, in every sector.

Of all the fields of government activity, defence is the first to have been able to organise itself with a view to mobilising technological progress : long before the beginning of the 7th century, while farmers in Europe attacked the soil with derisory wooden implements, the swordsmiths were forging the bright long swords which had defeated the Roman legions. In industrialised societies, however, defence has a broader meaning, and the concern for prestige and national independence very largely explains why technological progress has also been mobilised - beyond military ends themselves - in the service of agriculture, medicine, energy and space exploration. In a period of strains and international competition, such achievements were able to emerge, grow and become institutionalised with the help of the scientific communities and the support of the public, which did not question the aims pursued.

In the same favourable climate - in which the "international challenge" always played an important role - industrialised societies have also endeavoured to maintain and accelerate the race for economic growth : with very unequal results, since the contribution of science and technology depended closely on the ability of industrial and social structures to utilise them effectively. The development of firms exploiting new processes or distributing new products bears witness to the success of those structures which were flexible and diversified enough to respond to the demands of the pursuit of innovation. The development of certain multinational firms provides an example.

By modifying the conditions of technological competition, international détente was instrumental in causing many of the objectives which were taken for granted in earlier decades to be reappraised. The internal challenge took over from the challenge from outside. At the same time, however, concern for the environment or for the quality of urban life led to a redirection of scientific and technological efforts. Could not the methods adopted for the mobilisation of knowledge, which had succeeded so well in the past, be applied to other fields where serious problems call for quick solutions ? Success is not automatically assured.

For several years, however, the automatic nature of the transformation of knowledge into innovation had already been questioned. The specific character of the major nuclear, military and space technological programmes had to be acknowledged : fruitful as they were, there could be no question of transposing them as such to other fields since their success was largely due to the fact that they took place within hermetically sealed systems : the customer - in this case the public sector - dictated in advance the specifications of the products desired, shouldered the risks of research and guaranteed the market.

In fields such as health, regional development, social welfare, town planning, the environment, housing or transport, research has to be carried out in "open systems" : the absence of market mechanisms often makes a realistic assessment of demand and requirements difficult, and this is the case throughout a whole range of activities in which government responsibilities are increasing. In some cases, the development of new techniques may be facilitated by the existence of a market - as in the case of water purification plant - although the present inflexibility of the system often prevents such a market from acquiring its own internal dynamic. Usually, although they have the necessary resources to launch research programmes, governments lack reliable criteria for defining such programmes, and moreover have practically no means of ensuring that the results, if any, will be used. In matters relating to education, housing or health, the structure of supply shapes the demand structure, since the consumer cannot make any real choice and, for lack of information, is almost never able to express a preference. In such circumstances, there is nothing to ensure that the positive results of research will be incorporated in the practices of the technical specialists - particularly when they involve considerable changes in acquired habits and organisational methods in a trade or profession.

The situation varies a great deal from one country and one sector to another. However, it reflects the difficulties encountered by any society which wishes to encourage the transfer of technology from one sector to another, facilitate the dissemination of innovation from one social group to another or come to the aid of a particular minority.

It is not enough, therefore, to launch a research programme, even if it is specifically guided by the desire to solve a particular problem : the results will remain unused or will be only partially used, and perhaps in a harmful way, without the agreement of all the possible "customers". However, the latter have little opportunity for taking the initiative since it is their very fragmentation which is responsible for their lack of power. By increasing the power of the State and extending its influence over the whole of society, the association of knowledge with the power to use it has given the State a special responsibility. It is often alone able to undertake research which is in the public interest. It must, nevertheless, define this "public interest" : how can effective action be taken in such circumstances without excessively increasing the weight of the State in society, save by helping the groups concerned to associate and organise themselves with a view to stimulating the interest of the public authorities in their problems and to supervise the implementation of the programmes with which they are concerned ? It is nonetheless true that it will always be difficult to reconcile the desire for participation with the concern for efficiency, the interests of some with those of others and resistance to change with the desire for change.

A dynamic approach and the ability to change can base power on the exploitation of knowledge. However, as it becomes established, this success also represents a factor of resistance to change. It is only with some difficulty, and after delays that are often fraught with consequences, that research and development programmes finally take account - sometimes in a purely symbolic way - of considerations other than the immediate objectives assigned to them. In defence, and in industrial or medical development, technological progress seems to have become the essential aim of technological progress. There can be no question of denying the very positive progress made - such as the lengthening of average life expectancy - but it must be realised that such developments have occurred with no particular concern for the dislocations and inadequacies accompanying them. The recent development of social medicine in some countries bears witness, for example, to the concern to bridge the gap between the achievements of medical technology and the high cost and poor quality of the medical attention available to the public at large. Many specialists are disturbed, moreover, at the spectacular progress made by agricultural and food technologies , whose products do not take account of the advances made in dietetics. In sum, the fragmented approach to problems on which industrial societies have based a major part of their development leads to serious imbalances when it comes to applying knowledge, because such an approach helps to conceal the complexity of the problems.

This disequilibrium even ends by acting as acting as a brake on the utilisation ofknowledge in those fields which are best prepared for taking advantage of technological progress. For instance, everyone expects government authorities to be able to follow the social and economic situation closely in order to take timely measures to protect the citizens' security and prosperity - on condition that such social observation does not go too far and threaten the privacy of individuals. In particular, the possibilities afforded by computers have made it increasingly realised that, in the absence of proper safeguards, the relative inefficiency of government services is, in fact, a protection for the citizen.

It is probable that, in the future, data and privacy protection laws, especially in their physical security aspects, may well influence information technologies. This would then become one of the first cases where social control of technology has been established sufficiently early to prevent major abuses.

Nevertheless, for the moment, the existence of a technology does not ensure that it will be fully used when fundamental values are opposed to it. Science offers opportunities which society is powerless to grasp for lack of adequate means of control.

B. SCIENCE AND GROWING HOPES

Economic growth has made certain inequalities more bearable, but its adequacy to meet qualitative aspirations would seem to be constantly diminishing. Redistribution of the benefits and disadvantages of progress has already become one of the major problems which industrial societies will have to overcome, and it will no doubt have a decisive influence over the direction taken by science policy. Any economic or social change is a source of difficulties for certain groups in spite of the benefits which the community as a whole may derive from it. However, the origin of these difficulties is ever more clearly perceived by those concerned, who increasingly refuse to put up with them. Introduction of a process of change which has as little as possible traumatic effect on society has therefore become an essential prerequisite for change.

Technological progress probably does not create inequalities ; all it does is to make some of the inequalities which have always existed more acute : the improvement of farming techniques, for example, increases the gap between "small" and "big" farmers. Such changes are increasingly perceptible due to the social observation instruments already available, and are regarded as less and less consistent with the rise in collective and individual aspirations peculiar to the present period.

In all ages, certain classes and certain groups have been on the periphery of social life. In a technological society, certain exclusions tend to disappear while others become more pronounced. The market machinery and the vagueness of political objectives favour the individual use of technology rather than its collective use, as is demonstrated by the development of the motor car at the cost of public transport. A civilisation which is self-defined in terms of the motor car increases the often tragic isolation of those who do not have access to it. A civilisation in which communications systems play an increasing role debars from political, social and cultural life those who do not have a telephone or television set - and the development of cable television may accentuate this exclusion still more. Here again, purely technical solutions would probably not be lacking if the problems were not essentially social, economic and political.

Industrial society is constantly growing in complexity. Complicated legislation or regulations, the tangle of government departments and administrative procedures, and ever-increasing bureaucracy make overall perception of individual rights more and more difficult. The use of computers is a substantial factor for efficiency but it means that those in authority remain anonymous, and thus prevents any direct contact between the administrator and those he administers. The simple fact of living and defending one's rights and one's individuality is becoming a complex act requiring a degree of skill and a minimum acquaintance with procedures which are not always within everyone's grasp. For certain categories of individuals, the inability to live in a technological society takes on a truly tragic dimension because of this.

Knowledge has always been a source of power for those who possess it, but access to knowledge - when it really is of strategic importance - becomes more and more difficult as society becomes more complex. The techniques borrowed from the social sciences - opinion polls and surveys, information processing or the assembling of research date - give those who have access to them a power of prediction, and perhaps of manipulation or control, which is all the more disturbing as it is only rarely explicit. Whoever finances a research project concerning a particular social group has access to results which the "subjects" of this research have little means of interpreting. A society which does not give everyone concerned real equality of access to knowledge in many cases, in the last resort, sanctions *de facto* monopolies which are virtual examples of the "embezzlement of knowledge".

Growing awareness of the risks which the development of science and technology imposes on society as a whole, or on certain groups, increases the responsibility of the researcher. Increasingly, research on problems relating to the environment, health or the functioning of society provides arguments that have political or economic connotations. Research is no longer neutral when it leads to a nuclear or bacteriological weapon. Research is no longer neutral when it proposes pollution standards which could bring about the ruin of an economic activity. Research is no longer neutral when, despite the complexity of the task, it is supposed to determine the effects of a consumer product on the human organism. Research is no longer neutral when it challenges traditional conceptions of the individual's responsibility for his acts, for example by stressing the socially-induced nature of certain types of criminal behaviour.

The dubious nature of the neutrality of science is also demonstrated by the fact that the scientific basis of the conclusions it propounds is not always as obvious as it may seem. What degree of pollution is really dangerous ? How much is to be attributed to synergetic effects when deciding whether a food product should be prohibited ? To what degree is it really scientifically possible to extrapolate from the rat to man ? In describing the situation of a social group, what are the respective shares of scientific objectivity and the desire to defend a threatened minority ? A clear distinction is not always made when scientific results are presented between what is objectively proved and what is probable.

Formerly, only the social sciences were regarded as inspiring doubts of this nature. A growing number of scientists from all disciplines find themselves today involuntarily at the heart of society's conflicts and strains. Neutral science has been replaced by "captive science" and "critical science" - the nature of either one or the other being very often determined by their institutional location. Captive science can be mobilised without questioning the objectives assigned to it ; critical science uses the authority of knowledge to help challenge established practices. As a whole, science seems to be increasingly heterogeneous. But this heterogeneity serves a society which has to face complex problems whose diagnosis can but be uncertain.

The present unrest of industrial societies is sufficient to explain this development, which is all the more justified since it enables uncertified knowledge to be tested in the fire of controversies that reveal its limitations. But technological societies have based their future on the exploitation of reliable knowledge, verified by a process of scientific legitimation regarded as sacrosanct. Such societies no longer know how to use a science which is uncertain as to its ends.

C. SCIENCE AND FRAGILITY

The increased material well-being of industrial society is founded on specialisation and division of increasingly complex tasks, for which scientific and technological development has become a necessary mainspring. In doing so, however, it has made society more and more dependent on the efficient operation of machines or institutions. Having become increasingly interdependent, activities are so inter-linked that the failure of one can paralyse all the others : a stoppage in the postal services due to a breakdown in a relatively obscure service can, for example, paralyse an entire economy.

A "wildcat" strike by a small group of workers responsible for an activity on which others depend, threatens the equilibrium of the entire edifice. Technological society is therefore especially vulnerable to the consequences of industrial disputes which are to all appearances quite secondary. Social cohesion is therefore one of the essential conditions for the survival of industrial society.

In the years immediately after the Second World War, higher education expanded considerably in every industrialised country. This expansion was largely in response to the very high demand, expressed by the need for scientists, technicians and management specialists in rapidly growing economies. But economic necessities alone do not explain the massive effort made in every country on behalf of education. It had, in fact, become the expression of great hopes and faith.

The explosion in student numbers, which doubled or even tripled in nearly every industrialised country, is proof of this. For two decades, this explosion nourished the hope for the democratisation of education and indirectly - through more equality of opportunity of access to important administrative posts in government and industry - for a social cohesion, always desired but never attained. These hopes have been disappointed or at least shaken, for two complementary reasons which demonstrate that the university reflects social reality perhaps even more than it moulds it.

On the one hand, democratisation has proved very relative : access to higher education was first via entry into the middle classes and not the opposite, as had been long believed. Universities remained almost inaccessible to the children of manual or agricultural workers. On the other hand, the down-grading of degrees and students' difficulties in entering the labour maket raised doubts as to the ability of mass education to provide the equal opportunities sought.

In addition, the university proved slow to change, from either the standpoint of the organisation of education and research, which was unable to cope with the influx of students, or from the standpoint of the content of the education provided.

This resistance of the higher education is all the more serious since at the same time it gravely hampers the renewal and orientation of research themes.

In fact, for both education and research, the mass educational system has not yet been able to redefine its role in technological society. Does it provide vocational training ? A large part of this eludes it in favour of specialised institutions, or is even given direct "on the job" by the relevant firms or organisations. Does it give cultural training ? A large part of this also escapes it for the mass media which, no doubt more than education, forge the conscience of contemporary man. Does it lead to research and the creation of knowledge ? Here too, its once almost complete monopoly evades it, to the benefit of other public or private institutions.

The confusion and uncertainty of education are the confusion and uncertainty of society as a whole. No institution is above criticism ; the effectiveness and legitimacy of each and every one's function are being challenged. Contestation arises and increases at every level, becoming sufficiently strong to shake society, but powerless - in view of the very complexity of things - to propose convincing alternatives.

In addition, the malaise is all the more widespread since, in technological society, success can never be taken for granted. The rapid progress of knowledge in certain fields makes a man of 40 out-of-date if he has not been able to keep up with its pace. The development of technology since the beginning of the century has been such that it is increasingly attacking the very foundations of society. Values, laws and institutions fall before the onslaught of technology. Each tomorrow in each man's working civic and family life is uncertain. Of course, in a Welfare State failure no longer has tragic effects for the survival of the individuals concerned. But can the sentiment of social uselessness felt by the recipients of "transfers" become a permanent feature of society without adverse effects for the community as a whole ?

Science no longer holds out the promise of a golden age. In solving one problem, it appears to create a thousand others. A madman can halt air traffic in a capital city. A technical failure can plunge half a continent into darkness. The spread of an innovation increases the vulnerability of those benefiting from it ; a new cereal is perfected, but its adoption destroys traditional social structures and makes the economy more dependent on climatic conditions ; a new drug is produced, but it is at once necessary to find ways of overcoming its side-effects;a new technology is developed, but we have to think about containing unemployment and the repercussions on living conditions which may result.

The fragility of industrial societies is the price of their success : internal fragility, because of the chain of dislocations brought about by growth, and the introduction of innovations in ill-prepared ground ; external fragility, because of the multiplicity of the raw materials from which economic and technological development draws its substance : technological societies have set themselves up as models of development, without perceiving the limits which make universal development impossible without redistribution.

The rapid increase in the scope of governmental responsibility since the beginning of the century has led to the multiplication and extension of government departments. Bureaucratic structures are inherently rigid and adapt slowly to changing situations which call for quick decisions in response to complex collective demands. Greater transparency of decision-making procedures and more democratic participation at all levels are essential in order to ensure the utilisation of knowledge and to breathe new life into administrative machinery, which only too frequently takes refuge in long drawn-out procedures in order to evade problems that are too pressing and complicated.

Societies cannot avoid facing the problem raised by the good or harmful possibilities of science, as they travel the hard road towards inner unity. Science is not, of course, the only source of knowledge and it cannot provide a system of values or aesthetic standards. But it does propose a method of tackling and rationally debating problems, which seems to be essential at a time when communities are threatened with paralysis because of the certainties and the uncertainties involved in taking action.

Rationalisation of the social debate cannot remove the strains and conflicts which are, in fact, essential to the life of societies. But it can lead to a sharper perception of the stakes and the implications involved. It provides, therefore, the basic frame of reference for a society which wishes to base its future on full knowledge of the facts. To take into account the totality of scientific knowledge - from the natural to the social sciences and including the engineering sciences - is no easy matter. It implies the profound adjustment of social, economic or political institutions which have not learnt to take full advantage of knowledge.

III

THE SOCIAL FUNCTION OF SCIENCE AND TECHNOLOGY

Science, profoundly influenced by its new relationship with society ; and society, discovering or rediscovering its problems through the magnifying-glass of scientific and technological advance : this is the background against which science policies have to be formulated. The overall aims have been clear for a long time, but they are increasingly urgent : better mastery of technological progress, and a closer association between knowledge and the formulation and implementation of social policies. All this still remains to be achieved. Concrete tasks have to be specified and the foundations laid for a policy of knowledge, sufficiently ambitious for there to be a hope of associating science with all human activity, yet sufficiently realistic not to expect it to resolve all problems.

A. A POLICY FOR THE SOCIAL SCIENCES

It would be unthinkable to increase the social contributions of science and technology without a better understanding of individual and collective behaviour. Similarly, there can be no hope of achieving a better mastery of the social consequences of scientific and technological development without as precise a knowledge as possible of the mechanisms and methods of bringing about change in society. Lastly, we could not expect to master the process of producing and using knowledge more fully without understanding the functioning of scientific and technical enterprise.

This amounts to saying that science policy cannot take up the new challenge facing it without calling on the social sciences, learning how to better use them and to direct their development.

The social sciences resemble a universe with undefined borderlines and heterogeneous content. The disciplines included in the "social sciences", in the "human sciences" or in the "behavioural sciences" vary from one country to the other. In this respect, the observations already expressed in a previous OECD report are very illustrative.(1)

1) The Social Sciences and the Policies of Governments, OECD, Paris, 1966, pp. 21-22.

The term "social sciences" is used here to cover the complex of disciplines concerned with the behaviour of man in society and of social institutions. These include sociology, psychology, economics, political science, demography, social antropology, and the social aspects of legal, historical and geographical studies. Each of these disciplines is distinct in specialisation and character, and each has its intellectual and methodological approach to problems. All, however, make use of mathematical and statistical tools, and different types of model-building.

It is impossible to give a definitive listing of these disciplines, because their designations, their division into academic disciplines and the links between them vary from university to university and from country to country. According to national tradition and orientation, certain fields covered by social science may be the subject of specialised teaching and research in some countries but not in others ; for instance, industrial management or market research. Furthermore, the present division into academic disciplines or branches of research is constantly under revision to keep pace with growing specialisation.

National traditions alone do not account for the fact that terminology varies from country to country. More often the different classifications used correspond to differing conceptions and research orientations : for instance social anthropology, cultural anthropology and ethnology are by no means equivalent ; in some countries "social geography" embraces much of what is known elsewhere as sociology (as well as some aspects of regional or urban planning). Even the more general terms are ambiguous : social sciences, human sciences, behavioural sciences are understood differently in different countries, sometimes grouped together within university structures and sometimes scattered amongst the different faculties. This ambiguity is inevitable since these disciplines, because they deal with human beings, cannot be separated from the cultural background in which they operate ; they therefore share common ground and are closely inter-connected with the humanities.

1. Utilisation of the social sciences

The utilisation of the social sciences involves a deliberate process of the creation, and more especially transfer, of knowledge, while their applications are at the mercy of chance which may bring about the accidental meeting of a social demand and a possibility of using knowledge. Can the potential value of the social sciences be systematically explored ? Certainly not everything that is "usable" is in fact being used, and the process of social selection thus brought into play is the fruit of constant confrontation between preferences and necessities, between values and practical considerations. It thus lies at the centre of the tensions and conflicts which it reflects or aggravates.

Of what use are, or could be, the social sciences ? What kind of contribution can they make, and what are its limitations ? How can a policy be shaped which will increase that contribution where it is most needed ? A Secretariat study on the development and utilisation of the social sciences by Governments has tried to answer these questions.

There is no single market for knowledge, the use of which forms an extremely complex pattern. Within government institutions, for instance, the problems differ according to level and function. The knowledge emanating from the social sciences can be drawn on to lift a corner of the veil which hides the future, and so offer society the possibility of acting instead of reacting, of parrying shocks in advance rather than dealing with them as emergencies when they arrive. This knowledge makes it possible to clarify issues, state options clearly and reject objectives which cannot be achieved. At another level it makes it possible to proceed by a series of successive adjustments in order to keep the implementation of programmes closer to their targets, and to draft new regulations so as to take better account of their indirect effects. There can be descriptive studies to illustrate the obstacles encountered in implementing policies. In all, it might seem that the immediate practical value of the social sciences is perhaps in inverse proportion to the size and complexity of the political problems involved : they become more difficult to use, and more unreliable, as the weight of individual and collective constraints and preferences increases.

One should not however conclude from this that the usefulness of the social sciences is of a comparatively secondary nature. Admitted the duration and implications of an ambitious piece of research are often incompatible with the needs of day-to-day action by decision-makers. The social sciences frequently make it possible to deal with social realities in a radically new manner which calls basic assumptions into question and, as with the natural sciences, really fundamental innovations are not acceptable without long and difficult periods of transition. The influence of the social sciences on the basic choices of society is a long-term one, which makes itself felt indirectly by changing scales of values, mental attitudes, and individual and political behaviour. The history of legislation on abortion provides sufficient proof of this.

These considerations suffice to show that the problem of making use of the social sciences extends far outside government circles : the use of knowledge affects the whole of society, and cannot really be achieved except with the participation of all institutions and groups. If this is not forthcoming it is particularly difficult, and may even be impossible, for the community as a whole to "digest" knowledge, and this results in what amounts to a "misuse of knowledge" for the benefit of one or another interest group.

There is, therefore, no single market for knowledge. Knowledge itself is not homogeneous. The "products" which the social sciences offer to users are extremely varied. Movements of people constitute the most simple and general form of exchange : a sociologist associated in the preparation or implementation of a policy may be able to offer, in addition to his own scientific value, a body of experience acquired in the field in direct contact with the groups concerned. At the same time, in addition to the results of research, the social sciences can provide information techniques and methods of approaching problems : reports on past work and collections of data, systems analysis, simulation models, etc. When these techniques are used to assist the decision-making process, they become, in the strictest sense of the term, "social technologies".

The various categories of "products" of the social sciences do not form an automatic sequence. In particular, the interactions between pure and applied research, like those between the various types of applied research, are extremely complex. Thus pure research nourishes applied research by providing it with conceptual frameworks on which it can build. Applied research aids pure research by proposing new problems reflecting the impulses it receives from the demands of society. In this unbroken chain of activities extending from research to application, applied research is the interface between policy and science. For each type of policy demand there is a scientific requirement : in analysing, suggesting means of action or defining strategic objectives, theory plays an important part.

Often, demand can exceed the possibilities of science because there are only a few fields where the latter can build on overall theories, which are, incidentally, often disputed. As for the theories of complex systems, they have scarcely begun to be elaborated. From the point of view of application, some disciplines are clearly more developed than others. In consequence, users often tend to exaggerate or misunderstand the inadequacy of the overall development of the social sciences.

Every government administration is, however, virtually a user of the social sciences, either because its activity is directly concerned with social problems (welfare, employment, education), or because it has important social effects (transport, communications, technological policy), or because its function is to organise certain communities which have specific problems (defence) or because its work stems from a desire to mould and organise the future of the community (planning).

In the past, more substantial contributions were made indirectly, through the intervention of public opinion or political pressure, calling into question the very concept of certain programmes. This is an example of a genuine underground influence of the social sciences, gradually eroding the assumptions on which certain policies were based. Thus education or manpower policies have, over the decades, undergone a real process of dislocation whose effect has been to diversify policies by adapting them to the specific natures and problems of various groups or minorities.

Many administrative authorities have tried to organise their relations with the social sciences, so essential has it seemed to them to create an "institutional interface" where there could be more direct confrontation between their own concerns and the themes or conclusions of research. Some have gone even further and directly associated the social sciences with the elaboration of their programmes. A decisive step in this direction was undoubtedly taken with the launching of social experiments providing information for decision-makers before a final decision was taken, and at the same time enabling social scientists to have a more direct appreciation of social reality.

In practice, however, the results do not necessarily correspond to the expectations of those concerned. Administrators are disappointed to find that the conclusions of research are not of immediate use to them ; while social scientists are put out by finding that the limits set on their work are too narrow, and that their results appear to be neglected.

These disappointments are certainly due to numerous factors. In particular, the development of exchanges between decision-makers and scientists is too recent a phenomenon in the social sciences for it to rest on a tradition through which each side can acquire a realistic view. Decision-makers are tempted to see any research project dealing with a social problem as the key to a solution ; while social scientists are impatient at a political process whose different rationalising process often escapes them.

This association also raises new methodological problems. The time required for a university research project to mature is too long for the decision-maker, who presses the scientist to produce rapid results, and is often obliged to take a decision before the research is finished.

Many of the difficulties involved result from the very nature of the relationship between the administration and the research milieu. No administration wishes to be directly called into question on account of a research project which it has inspired ; while few social scientists are prepared to compromise their relationship with an administration which is financing their work. The closeness of this "capacity" of social science undoubtedly varies (a research scientist in a government institution is usually more captive than a university research scientist under contract who, theoretically at least, has access to other sources of finance), but it does enable the sponsor to set limits on the research which prevent it from extending into dangerous territory and may even influence the results.

In these circumstances, a policy for utilising the social sciences still has to be evolved step by step in the light of everyday experience. Some certainties are already beginning to appear. Such a policy would indeed have to be concerned more with facilitating the transfer of knowledge, and enquiries within society, than with increasing the number of institutions, each acting as a filter. It would have to try to develop techniques for the better deployment of research scientists, as dictated by the main concerns of the moment (e.g. the constitution of genuine inter-disciplinary groups in which certain more "qualitative" disciplines would not be eclipsed by those that appeared to be more exact ; the evaluation of programmes ; experiments in the social field serving as demonstration projects or as studies preliminary to the launching of a programme ; etc.). The policy would also have to go beyond the framework of government institutions, and stimulate interest in the social sciences at all levels of society.

A new policy for knowledge must in fact try to bring about a redistribution of knowledge, which alone can endure a revival of the social debate through democratic participation. A policy for the use of knowledge is also one of incentives to make use of it, just as a technological policy is a policy of incentives and guidance to make use of technology.

This rules out any approaches that are too exclusively dictated by immediate needs. The contributions of the social sciences depend on their ability to extend and deepen the area of investigation. A policy of utilisation cannot be dissociated from a policy for the development of the social sciences, designed in particular to intensify exchanges among all the disciplines of knowledge. It is therefore a matter of bringing the production of knowledge back into balance.

2. A policy for the development of social sciences

It is impossible to make wider use of the social sciences without first having basic knowledge of the various fields in which they are applied. Resort to model-building, to so-called advanced techniques, or to increasing the number of studies and surveys, can only be a substitute for a lack of basic knowledge of social phenomena. The possibility of producing descriptive information (numbers of unemployed numbers of poor people, distribution of social deviants) depends in the last analysis on the value of the reference categories adopted to classify such data. The defects of scientific theory become evident when the various items of information have to be confronted, compared or integrated in a wider field.

It would be equally unwise to rely on social technology or social engineering based on objectively observed phenomena without understanding the laws governing them, in the way that experimental medicine has been able to cure diseases by vaccines even when the nature of the diseases was unknown. The social data are far more numerous, complex and changing and their evolution defies the analytical ability of the most powerful computer left to its own devices.

Greater and better use of the social sciences should not, however, be sought independently of the other areas of knowledge. The complex interaction of all social phenomena, and the essential part played by technology in their development, make it henceforth impossible to conceive an evolution of the social sciences dissociated from the natural sciences. The effort required to achieve better integration, and associate the social sciences with the life of the scientific community, should above all be an occasion for profound reflection on the adjustments needed to the latter to make it more sensitive to social impulses and better able to respond to them.

The absence of institutionalisation, which is the placing of a field of study within an organisational framework, is much more apparent in the social sciences than in the natural sciences. Fragmentation, isolation and dispersal of research, as well as the lack of a solid conceptual framework, are prejudicial to their sustained growth, the progressive ramification of their various branches, their rational financing ; in short, to the credibility of the whole system. Institutionalisation takes place mainly in universities or in government departments. Whereas the tradition in the United States is that departments and government agencies rely on universities to carry out the research they need, under contract, European ministries have often been obliged to develop their own research capacity often via the departments responsible for supplying them with information and statistics. Such services have gradually become involved in research in order to improve their statistical methods or their understanding of an area with which their parent administration is concerned. Alongside these departmental institutions, research institutions of a quasi-university type have also grown up, which may be made responsible for dealing with an important and chronic social problem such as alcoholism. These institutions are often smaller and function on university lines (freedom of choice of research topics, unrestricted publication of results), but the government departmental institutions are usually bigger and are subjected to specific requirements dictated by their department (working to order, confidentiality of data and results).

This increasingly complex network of government department centres has developed as a counterbalance to the universities, whose structures and aptitude for empirical research did not appear to be satisfactory. This solution, attractive in the short term, nonetheless has serious dangers in the longer run. The social sciences may not be represented in a balanced fashion in the technical ministries, and research work tends toward simple studies and surveys and lacks the methodological and conceptual support which would give it meaning. The range of research is often limited to the administrative responsibilities of the ministry concerned, as laid down by systems of government organisation, which are always arbitrary. The confidential nature of the research makes it impossible for the scientific community to assess it accurately. Finally, the quasi-university type institutions which are often set up to meet a politically important need may "freeze" the definition of the problems entrusted to them, and work in isolation.

All these factors lead us to recommend more systematic resort to the universities in order to develop research. The university is perhaps less captive, and has the independence needed to criticise the conceptual and practical programmes of the Administration.

Such privileged financing of university research in the social sciences cannot, however, bear fruit except to the extent that the scientific community itself - including the natural as well as the social sciences - becomes better adapted to present-day needs. Some reflection on this matter is essential in order to avoid making ill-considered or hasty changes. In particular, an attempt must be made to improve career structures, rewards and sanctions, as well as training methods, so as to make research better fitted for tackling social problems and to ensure a continuous link between science and the application of knowledge.

These general problems should not, however, obscure the fact that the situation in the social sciences at the universities is still far from displaying the trend already begun in the natural sciences. The universities do not yet provide the solid foundation needed for research in the social sciences. Their development there is comparatively recent, and generally faces severe constraints : conflict with the demands of teaching, difficulties of inclusion in a disciplinary framework, and a lack of material resources. The division of the social sciences, dispersed among a host of small institutions, is too reminiscent of the time when research in the natural sciences was exclusively the concern of professors, appointed to a particular chair, with a few assistants, devoted both to the pursuit of knowledge and to petty parochial squabbling over schools and doctrines.

Increasing the research potential means correcting any imbalances that may have risen in the universities : reducing the teaching load where it is too heavy and encouraging the growth of new means of research. In this connection information technology cannot be regarded as the sole instrument of research. In particular, social experimentation can make a considerable contribution to research without being excessively costly, as has been shown by the experiments in education which have been widespread since the early 1930s.

The quality of research is difficult to determine, but it is not impossible to find ways of measuring it. Clearly research needs time to carry out and to develop, and the funds allocated to it must take account of this. The budget for the social sciences is growing considerably, but its structure shows that the proportion allotted to research councils is often substantially less than that available to government departments and through them for so-called contractual research. In these conditions, the research institutions are diverted from pure research, either because they do not have the means to pursue it, in the form of subsidies from research councils, or because they do not have the time because of the strict requirements of a contract.

We must not, however, overlook the fact that a policy for the development of the social sciences at a time of comparative scarcity of resources cannot avoid making a choice : the field that can be covered by research is unlimited. How can reasonable and desirable limits be laid down ? A more flexible and continuing link between research and the demands of society is of itself helpful in defining what directions are to be preferred. A policy for development, however, must define the criteria for its choices more precisely, the more so as that policy is made indispensable by the distortions that may be introduced into research by too much insistence upon short-term social, economic or political considerations.

The attempts made in several countries to evaluate for each discipline the state of knowledge and the needs of research have given some indication of the unbalanced state of scientific development. These studies do not, however, enable us to draw up a real list of priorities, and thus they are of limited value insofar as they do not deal with the relationship of the various disciplines with one another.

At a time when it is clearly more urgent than ever to adopt an interdisciplinary approach, bringing together specialists of the natural sciences, engineering science and the social sciences, the most delicate problems are those posed by the relations between such widely varying disciplines. The danger that one discipline will dominate, or spread over into another is especially disquieting at a time when those using knowledge have not yet learned to take certain forms of knowledge into account. How can we prevent quantitative approaches from driving out qualitative approaches that appear less exact and are less easy to use, but perhaps reflect better the complexity of social phenomena ? The problem arises not only in the framework of the relationship between the social sciences and the natural sciences but within the social sciences themselves. The hearing and recognition that politicians and administrators accord to economists, for example, have steered the teams responsible for evaluating government programmes in a direction which overlooks the sociological and political realities which should be taken into account : the value and credibility of such evaluations are profoundly affected by this.

Conversely, the intellectual authority acquired by certain disciplines is harmful to their own development in that it prevents them from taking into account other aspects of knowledge. Thus, the contributions of psychology and sociology to economics are still extremely limited : economics, before the other social science disciplines, has become established in its own right and as a profession ; its very individual structure has perhaps limited subsequent opportunities for other disciplines to contribute to it. The social dimension of scientific and technological development cannot be truly appreciated without a constant confrontation of disciplines one with another.

B. GETTING THE TECHNOLOGY WE WANT

The present is increasingly overshadowed by the future. Technological developments as yet only planned or foreseen as possible already play an important role in political and economic life : future prospects relating to exploitation of the sea and seabed, to the methods of production of raw materials, or to bringing new energy sources within our reach, already lie at the core of many international negotiations. The progress made with instruments for the observation of society is also leading society itself to seek to determine its main options in the light of forecasts of trends in economic and social activities.

Scientific and technological development lies at the very hub of all these concerns : it may assist in the refinement of forecasting techniques but it is also itself the subject of forecasts. The future of society is too dependent on it not to try to foresee and even to influence its course.

Clearly, in the fact of the possible dangers and new problems arising from or brought to light by, the development of technology, change does not imply merely improvement in social well-being and the satisfaction of collective and individual aspirations. The traditional criteria of economic growth seem to be of limited value and accordingly the reliability of the indicators used to measure it is increasingly questioned. This is because they are seen as concealing the complexity of the problems involved by providing statistical aggregates which are difficult to interpret but which end - like the gross national product - by acquiring an independent existence, as targets to be pursued, without any question as to their meaning. Another reason is, that such indicators are frequently accused of amplifying the facts in order to be able to measure them, and thus "quantifying the unquantifiable". Lastly, they are called into question precisely because they are unable to take into account essential qualitative aspects.

This suggests that science and technology policy will be called upon to invent a new form of dialogue between society and technology, which will take into account man's aspirations in a complex society, rather than the "autonomous" trends of science or technology. If lethal disasters and socio-cultural collapse are to be avoided, better means of ordering the systems devised and built by men must be found, and scientific research and technological innovation must be better directed to achieving social progress.

The fundamental issue is, to what end is technology directed ? Have we not lost sight of this end, because of the fact that both scientists and the general public firmly believed that technology was a good in itself, the result of an autonomous process ? Yet there is no such thing as a scientific or technological objective in the sense of the application of a process of evaluation, the assignment of a certain value. This is essentially a political question.

The conclusion to be drawn from this is that the notion of social innovation set in motion by technology is a contradiction in terms. There can be no genuine social innovation resulting from a simple adjustment to an autonomous technological development. This is a false conception of the interrelationship between society and technology. If this conception is nonetheless very widely held, it is probably because it is a convenient way of unloading social responsibility on to an external factor.

How can the social management of technology be conceived ? First, as a task of evaluation which will examine and take account of the direct and indirect effects of technology on society.

The term "technology assessment" would seem to have been used for the first time in 1966 by the Science and Research Development Sub-Committee of the Committee of Science and Astronautics of the United States House of Representatives. According to Emilio Q. Daddario, its chairman at the time, technology assessment is a form of research which places a balanced assessment before those responsible for decision-making. Ideally, it is a system which can ask the right questions, and produce the right answers in due course. It identifies the points at issue, determines the implications of various lines of action and reveals options. It is a method of analysis which systematically assesses the nature, significance, scope and value of a technological change.

As the idea has spread, the term technology assessment has begun to be applied to a very wide variety of activities, ranging from technical feasibility studies to cost-benefit analyses and technological forecasting. The technicians have tended to see this as a mere extension of systems analysis, essentially designed to give a broader justification of one or another new techniques.

Now, the really fundamental change consists in the idea that technology cannot be left to itself, but that governments - or better, all the various categories of decision-makers - should assume responsibility for its development. This leads to the concept of a social evaluation of technology, which implies that the approach should be one which enables the underlying strategies and actions - of whatever origin - to be compared in the light of their total effect on society. It should help to make the process of political decision-making more explicit, by ensuring that the general interests of the community, as well as of the individual, are taken into account as far as humanly possible.

Two further ideas flow from this : firstly, that of a suitable time-horizon, and secondly, that of a range of identifiable consequences. The former points to the need for an "early-warning" system which can spotlight sufficiently in advance the point of transition from scientific results to technological innovation. Such a system should show the benefits and disadvantages of any important technological change, with a view to establishing technological priorities and suggesting new directions for the allocation of scientific resources. The second assumes that a mechanism exists for discerning the indirect effects and consequences, confronting these with direct and immediate benefits and putting the question of social relevance at each stage.

The fundamental difference between a technical and a social evaluation of technology lies in the scope of the analysis, its systemic (overall) nature, and the explicit treatment of social costs and benefits. This is a difference in kind. In the first case, the social aspects are either played down or expressly disregarded. In the second case, the prospective and evaluative aspects have to be fitted into an examination of the effects and repercussions on all the constituent parts of society.

As an illustration, we may take transport. A social evaluation of transport technology does not see transport merely as a means of moving passengers and goods from one place to another, but is concerned rather with the effects of present and potential techniques on housing and urban development, production, and the social and cultural functions of the various modes of transport. It is also concerned with communications of all kinds, with the cohesion or instability of communities, with the environment and, finally, with the conflicting interests of the various economic and social groups.

However, assessing the incidence of a technology on society is undoubtedly the most difficult and complex part of the process of assessment, because of the many unknown quantities and the basically uncertain nature of the phenomena of social change. These, unlike physical phenomena, rarely follow a regular or reproduceable pattern ; they are the result of irregular groupings of complex influences, which can act either as an accelerator or as a brake.

The purpose of social evaluation of technology is to assess technology in relation to what is socially desirable or expedient. Such an assessment presupposes a set of values on the basis of which certain estimations can be made about the socially beneficial aspects of technology. There are, however, many sets of values in society, frequently conflicting and often implicit. A minimum degree of consensus makes it possible to provide a social framework within which they can come into contact with each other. Outside such consensus compromises are essential if there is to be any possibility of defining lines of action. Hence, the social assessment of technology calls for collective thinking which can only become possible through genuine democratic participation. Hence also, the social assessment of technology needs to draw on a whole range of disciplines and to construct its own method as a dynamic blend of the natural sciences, engineering science, and the social and human sciences.

The social acceptance of a new technology can vary between opposing it on principle, or accepting it without reservation according to whether the parties concerned perceive it as threatening or beneficial to their style of life. These attitudes are not always based on the real threats or promises, but result from the inability to comprehend all the implications of a technological development.

Each profession or group thus has blinkers on which prevent it from relating its own problems to those of society, or conversely, from taking into account general trends in the community in determining its own future. This cannot be attributed to the specialisation of work or to competition, either of which may bring one interest, sometimes brutally, into conflict with another. The way in which knowledge is distributed throughout society is in itself a considerable obstacle to a broader approach to problems.

Sometimes education helps to perpetuate a literary and philosophical culture for whose proponents technology is a vulgar, material preoccupation. At other times or places, education maintains the tradition of a scientific and technical culture developing according to its own logic and consequently not concerned with the great moral, intellectual or social currents which motivate society. Thus, two radically different cultures exist side by side, and sometimes even co-habit in the same educational institutions, without ever meeting or combining with one another.

Within such a system of transmission of knowledge which fragments culture and perpetuates old divisions, the difficulties arise which constitute obstacles to a symbiosis of the natural engineering, and the social and human sciences. The only way of removing these obstacles is to achieve a closer and more dynamic association between teaching, research and society.

C. SCIENCE AND SOCIAL POLICY

The scope of science policies has so far been confined to a comparatively small area of government activity that was most directly concerned with technological development. It is only recently that, beyond questions of defence, energy or allocation of resources which had been their main concern, these policies have been extended to public services such as transport, communications or housing. The new factor was that in these fields, the problems could be stated only in social, economic and political terms. Together with the need to take the effects of technological progress into account more generally, this development introduced a social dimension into science policies.

The problem of the social sciences was thus raised. At one and the same time the actual mechanism of the alliance between science and policy-making which developed immediately after the Second World War, was called into question. Should not this alliance, which was mainly concerned with technological innovation, also take social innovation into account ? Should it not extend to social policy, where technological research plays a smaller part that social research ? In a word, it became clear that the whole history of the development of policy for science and technology in recent decades was a kind of prologue to the institution of a more general and systematic association of power with knowledge. This development gives rise to many difficulties and much opposition, as it calls into question both institutions, structures and traditions. The basic nature of national scientific and technological undertakings is profoundly affected by it.

At a time when the main scientific and technological effort was concentrated on the big strategic military, nuclear and space programmes, it enjoyed the advantages of being very stable and highly specific : its goals, and the means of attaining them, could be clearly defined in technological terms from the very beginning, and were mainly confined to specific projects such as the development of a military device, an aircraft or a moon landing. In the main, the setting of such objectives was identified with a given scientific and technical development ; and it was therefore possible to draw up a timetable for carrying out such technological projects as soon as the objectives were defined.

Socio-economic objectives are not amenable to such a approach. The improvement of the conditions of life in towns, the reduction of crime, or provision of adequate social and medical care, are all "moving targets", which involve continuing progress but are not in themselves problems which can be resolved once and for all. As a result, research programmes for this type of activity always remain by their very nature unfinished. They can evolve only by successive adjustments to the new conditions created by the progress of knowledge, by the trend of the problems, and by changes in the balance of the prevailing political, social and economic forces.

The former comparatively broad and stable goals are thus tending to be replaced by a whole range of "targets" - such as different social norms - that are both partial and changing. These "moving targets" to which research efforts have increasingly to be directed, require a set of converging projects rather than large-scale programmes. They also call for forms of organisation that are as flexible - and hence decentralised - as possible, to enable the necessary adjustments to be made.

The very nature of these problems thus calls for joint efforts by administrative organisations and regional and local authorities, together with non-governmental economic and social institutions. The latter have not hitherto been very much involved in the preparation or implementation of science policies, and it is therefore not surprising that they have so far been only partially successful in this work. Several Member countries are trying to launch large-scale national programmes which would enable the principal political, administrative, social, economic, scientific and technological interests concerned to become associated in joint tasks. Conceived in such a way, these programmes would involve a <u>concerted</u> mobilisation of research potential rather than a centralised organisation of the type which we have known in the past. Despite the creation of horizontal ministries in, for instance, the fields of environment, transport and urban affairs, these large integrated programmes have, in many cases, failed to fulfil the hopes placed in them.

Thus, the traditional conceptions of management by objectives are encountering considerable difficulties in practice, connected with the provisional nature of the objectives themselves and with the obstacles to their achievement ; while the definition of the targets and of the means, present a further challenge, which concerns the whole of society. In the present state of affairs, however, the two tasks are so closely complementary that it would appear difficult to make a distinction between them. In fact contrary to what traditional logic might suggest, the fixing of the means seems here to dictate the fixing of the ends. Any social policy is founded on a minimum of consensus among those concerned. Knowledge cannot hope to abolish conflicts and tensions, but it may define the limits within which rational debate is possible.

CONCLUSION

The co-existence of spectacular technological achievements and unsatisfied collective needs, the pervading impression of the dangers inherent in the present inequitable distribution of the benefits and disbenefits of technological progress, a technological development which broadens the range of options open to decision-makers, but also increases the possibilities for error : such are the fundamental problems with which science policy is faced in the exercise of its political and social responsibilities.

Science and technology have nourished the growth which has resolved or attenuated some problems, accentuated some others or even transformed them to crisis proportions. Threatened by new imbalances, industrialised societies are now seeking to initiate a process of evolution and change which will be better controlled and take more fully into account the complex of collective goals. This can only be achieved by learning to utilise more fully the knowledge provided by the natural and engineering sciences and the social sciences. Better ways must also be found to stimulate the production of knowledge which is lacking.

In many respects the inadaptations or resistance within the research system are a reflection of those to be found within society : the perpetuation, in one area or another, of traditional behaviour or structures ill-suited to respond to social demand ; compartmentalisation of research which prevents the multidisciplinary nature of social, economic and political problems from being properly taken into account ; dilution of responsibilities which limits the possibility of anticipating the consequences of scientific and technological development.

Apart from the malaise felt by both research workers and users of knowledge with regard to its utilisation or non-utilisation, there are new tasks to be fulfilled : a better exploitation of knowledge ; closer association and a better balance between the natural and social sciences ; heightened awareness of the responsibilities of scientists ; an effort to adapt existing structures to cope with new problems in a period of rapid change and increasing uncertainty.

Part Two

SCIENCE POLICY AND THE MANAGEMENT OF NATURAL RESOURCES

"In becoming concerned with energy and raw materials, economics policy is returning to the origins of political economy. The founding fathers of economics indeed never failed to emphasize the role of what they called "nature" or "the earth". But as from the middle of the 19th century, this voice was no longer heard. Economists, and the politicians who heeded what they had to say, were to reduce all problems to a combination, or confrontation, of labour and capital. Developments on the energy and raw materials markets now again compel us to take account of natural resources in our thinking and actions".(1)

The present concern with natural resources does not result only from the events which have affected the energy and raw materials markets over the last two years. It is also linked to the recent proliferation of studies and discussions concerning the long-term problems generated by the continued exponential growth marking recent decades. This line of thought culminated in the report "The Limits to Growth", prepared for the Club of Rome by a team from the Massachusetts Institute of Technology.(2). However open to challenge this report's methods of analysis and conclusions may be, it makes at least one fundamental point : <u>that the essentially finite nature of our planet and its resources is incompatible with continuous exponential demographic and economic growth</u>.

To this obvious, though too often forgotten, fact, must at once be added the qualification that, on the basis of present knowledge, there is little evidence of a world shortage of natural resources, many of which have yet to be discovered. If the controversy concerning the limits to growth has had such an impact it is because it coincided with a series of problems and crises affecting all types of natural resources - energy, materials, food, the environment - which afforded concrete proof of their fundamental importance to the economy, the daily life and even the survival of all countries, whether industrialised or developing.

Although the oil crisis has had the most conspicuous effects on world economic and political equilibrium, it must be viewed in conjunction with abrupt fluctuations in the prices of certain raw materials (for example phosphates or paper pulp), the new awareness of the

---

1) Speech by Mr. Emile van Lennep, Secretary-General of the OECD, at the Symposium of Energy and Raw Materials, Paris, 6th June, 1974.

2) Meadows, Randers and Behrens : <u>The Limits to Growth</u>, Universe Books, New York, 1972.

dangers arising from environmental damage and the shortage of food affecting nearly half humanity.

These difficulties and crises also showed that the nature of the problem of natural resources is primarily political, economic, social and technological. The 1973 oil crisis was mainly political in origin even if the prospects of the exhaustion of certain oil reserves may have been a factor in the decisions taken by some producing countries. From the economic point of view, while the industrialised countries are now faced with the consequences of an abrupt increase in the price of certain resources, it should be remembered that the economic development of many Third World countries had previously been adversely affected by deteriorating prices on the world market of most raw materials which were their main source of income. From a social standpoint, one leading reason for the vulnerability of industrialised economies is that their growth has been marked among other things by a process of rapid obsolescence and the replacement of still usable goods, i.e. the deliberate squandering of resources. From a technological aspect, the earth receives each year an amount of energy from the sun representing about 25,000 times that consumed annually throughout the world, yet we do not possess the appropriate technology to utilise solar energy on a large scale under viable economic conditions.

In other words, the main problem is not so much the physical limits to the planet's stock of natural resources as the inequalities and deficiencies which are a feature of the exploitation and use of these resources by society. Better resource management is therefore an absolute requisite.

There can be no question here of embarking on an exhaustive analysis of the whole subject of natural resources. Whatever the importance of science and technology, their role is limited, and the economic, political and social aspects are at least as decisive. Nor is the purpose of this paper to review, or even to outline, the specifically scientific and technological aspects. Its more modest aim is to identify, with a view to discussion by the Science Ministers of OECD countries, the main problems with which science and technological policies are faced in ensuring that research and development contributes to the better management of resources. In particular, an attempt has been made to apply the lessons learned from the work of the Committee for Scientific and Technological Policy in the fields of energy (1) and materials to natural resources as a whole. Before tackling these problems of scientific and technological policy, it was thought

---

1) Energy R&D - Problems and Perspectives, OECD, Paris, 1975.

advisable to describe briefly the main features of the natural resources field and its essential characteristic - the fact that it is a complex system.

I

NATURAL RESOURCES : A COMPLEX SYSTEM

Until very recently R&D related to the various types of natural resources, and to the different technolgies involved were considered independently of each other. The recognition of environmental problems, the increasingly acute shortage of food resources, concern with respect to material resources and the recent preoccupation with energy problems have clearly shown this kind of approach to be inadequate in a situation where the links between resources and between technologies are seen to be increasingly complex. Two concepts are deserving of special attention : the interdependence of the different kinds of resource and the life cycle of the resources themselves.

A. THE INTERDEPENDENCE OF RESOURCES

The interdependence of the various natural resources can be analysed at a number of different levels, first, that of the different categories of resource : energy, materials, food and environment.

Certain natural resources are sources of both energy and materials. One example is petroleum ; others are hydrogen and methanol which, although now used as raw materials by the chemicals industry, could well be utilised as energy carriers. Secondly, the processes of extracting, converting, utilising and recycling materials all consume energy : a fall in the grade of deposits being worked means that more energy will be needed to extract a given quantity of usable materials ; substituting other materials for a given application can appreciably change the amount of energy needed for the purpose ; if the lifetime of the material is longer, the total amount of energy needed for the material to perform its function will usually be smaller.

Conversely, there can be no energy production, conversion, transport or utilisation without materials. A recent survey of the prospects opened up by energy research shows how critically the development of all energy technologies depends on that of suitable materials.(1)

---

1) See for example Materials Technology in the Near-Term Energy Program, Report of the ad hoc Committee on Critical Materials Technology in the Energy Program to the Science Adviser to the President, National Academy of Sciences, Washington, D.C., 1974.

The factors linking energy or materials with food resources, although less immediately obvious, are equally fundamental. It will be remembered that petroleum can be converted into single-cell food proteins. The basic point is that agriculture and, more generally, the food system, consume substantial amounts of energy and materials. While agriculture is basically the organised utilisation of solar energy by plants, it requires an additional consumption of energy,(1) at present largely obtained from fossil fuels. The importance of phosphorus and potash for fertilisers may also be mentioned. Conversely, organic wastes, the major part of which comes from the food system, can constitute a source of energy through combustion, pyrolysis or biological conversion.

The factors linking energy, materials or food resources to the environment are even more complex. These may be regarded first from the pollution standpoint. Certain types of pollution are directly related to a particular type of resource. Thus in the case of energy, carbon-dioxide emissions inherent in the use of fossil fuels, or the thermal pollution associated with any type of energy consumption ;(2) in the case of materials, mining spoil and chemical pollution associated with mercury, cadmium, lead and other heavy metals ; and in agriculture the pollution resulting from the use of chemical pesticides. Moreover certain types of pollution are not directly linked to one particular resource but rather to several combined in various technologies (as in the chemicals industry).

Any investigation of pollution problems must also take account of the following factors : in the first place, the geographical distribution of pollution is quite as important as its volume ; nor is there yet any satisfactory method for comparing the different types of pollution (e.g. the emission of sulphur oxides with radioactive waste) ; finally, when several technologies can be employed to serve the same purpose, the one best from an environmental standpoint does not necessarily consume the least amount of energy or materials.

The relationships of energy, materials or food resources with the environment cannot however be viewed only in the context of pollution. The production, conversion, transport, storage and utilisation of energy, materials or food resources also require resources such as water or land, which are limited in extent, and variable in quality,

---

1) In the United States, such additional consumption during 1970 was estimated to be 3.8 per cent of total final energy consumption ; this figure is 15.6 per cent if the entire food system is considered(agriculture, food industry, storage and the preparation of foodstuffs in households). See John S. Steinhart and Carol E. Steinhart, "Energy Use in the United States Food System", Science, 19th April, 1974.

2) With the exception of solar energy.

unevenly distributed and also needed for other human activities. For instance water is important for electricity generation, irrigation, drinking and recreational pursuits. Another example is provided by problems of land-use caused by open-cast mining and the siting of power stations.

To the interdependency of the relations between energy, materials, food resources and the environment must be added those among different resources of a given type. A noteworthy example is the competition between the various energy sources (oil, natural gas, coal, nuclear energy, solar energy, etc.) or between various materials which can be used for the same function (such as stone or concrete and steel as building materials, natural and synthetic rubber, copper and aluminium as conductors of electricity). In fact a resource is valuable only by virtue of the properties fitting it for a particular function, and seldom can this function be fulfilled by only one resource. In this respect, air and water of course are a special case because of their fundamental importance for supporting life.

The possibilities of substituting one resource for another are therefore of capital importance since they provide a way of overcoming the constraints set by the relative scarcity of some particular element. Substitution should not however be viewed merely as the replacement of one resource by another for some given application : substitution of products (private car/public transport) or functional substitution (transport/telecommunications) hold out the prospect of a much more flexible use of natural resources to meet man's needs.

The different energy sources and materials resources do not always compete but may also complement each other. For example, heat from high-temperature nuculear reactors can be used for the gasification of coal, and in the case of materials, the increasing importance of alloys, composite materials and the fundamental role of catalysts in the chemicals industry may be mentioned.

B. THE LIFE CYCLE OF RESOURCES

The interdependence of natural resources only takes on its full meaning in consideration of their life cycle, starting with the resource in the raw state and proceeding through all the stages of conversion and use, up to and including the events which follow final utilisation.

The concept of a life cycle has long been familiar in the case of food resources, partly because of the close and evident link between agriculture and the changing seasons. It was in fact the scientific study of plants which revealed the existence and importance of the water and nitrogen cycles in the atmosphere and in the soil.

Only recently was it realised that the life-cycle concept also applies to energy and materials. In the case of energy, the cycle is an open one, since any consumption of energy ultimately results in heat being given up to the atmosphere. Non-renewable energy resources (fossil fuels, nuclear resources) must however be distinguished from the various forms (direct or indirect) of solar energy, of which considerable (but finite) amounts steadily reach the earth in the form of electromagnetic radiation.

In the case of materials, however, the cycle is closed.(1) There is no true consumption, instead a process of conversion and diffusion of resources remaining on the earth occurs. For this reason special importance should be given to the recycling of materials. In other words, wastes should be regarded as resources which compete with and complement raw materials.(2) In addition, a distinction should moreover be made between those materials resources which are non-renewable, (such as minerals) the recycling of which calls for human intervention, and renewable resources (such as wood), which are recycled through a natural process,(3) largely dependent on solar energy.

Apart from such differences in life cycles, the next most important point is that the various stages or technologies making up the cycles of a resource all interact with each other. The technical, economic or other problems arising at any stage of the process can be fully grasped only if all phases preceding and following the technology in question are taken into account.

For instance, problems of nuclear energy, whether the economic or industrial aspects, environmental effects or safety questions cannot be properly evaluated by considering the nuclear reactor alone. They must be seen in a wider context, comprising the whole nuclear fuel cycle (including radioactive wastes) as well as the entire energy cycle covering the transmission, storage and use of the electricity produced.

To take another example, the use of energy in industry should not be considered merely in terms of energy consumption for each separate technology, but should be approached by studying the total amount of energy consumed in the whole process extending from the raw-materials stage to the finished product, including its utilisation and any recycling.

Because of the number and complexity of the relationships outlined above, the field of natural resources cannot be regarded simply

1) Disregarding helium.

2) For this reason wastes are sometimes called "secondary resources".

3) This process can of course be influenced positively or negatively by human intervention.

as a number of co-existing resources and individual technologies, but as a veritable system. In particular, it is at the level of the overall system that the various factors (later to be discussed in Chapter II) to be taken into account in determining the main lines of research must be evaluated.

C. THE SOCIO-ECONOMIC DIMENSION OF THE NATURAL RESOURCES SYSTEM

As stated in the introduction to this paper, the natural resources system cannot be reduced solely to the scientific and technological aspects listed above. It can be fully understood only if account is taken of the relationships linking natural resources and their management to the economic and social framework in which the resources are recovered, processed and used and in which the relevant technologies are developed and applied.

All societies are dependent upon natural resources since they are essential to man's survival, but the ways in which they are exploited and utilised have a profound influence on societies, their modes of life and structures. It is not without significance that prehistoric civilisations should be known by the names of the main materials they used, e.g., the Stone, Bronze and Iron Ages. Mention may also be made of the decisive influence exerted by the "invention" of agriculture, which can be regarded as the organised utilisation of solar energy through plants on the sedentation process. Yet another example is the predominant role of wells in nomadic societies.

More recently, the development of synthetic polymers, the use of coal in the steam engine, of oil in the internal combustion engine and the harnessing of electricity have given rise to completely new industries, and radically affected not only the economy but also the structure and development of industrial societies. Similarly, the trend towards an agricultural industry based on the increasingly intensive use of energy has decisively influenced the growing urbanisation of these societies in the last few decades. Today, the development of nuclear energy and such consequences as the increasing amounts of long-lived radioactive wastes, place entirely new responsibilities on these societies with regard to the world as a whole as well as to future generations. These responsibilities mean that civilisations and ways of thinking must change.

Conversely, the type of civilisation and socio-economic structure does much to determine what kind of resources are used by societies and how. This influence is exerted in many and varied ways, for example through science and technology. For although the scientific basis of technologies is independent of social organisation, the selection of subjects for R&D, the choice of technologies, research capacity and the ability to apply the results do in fact depend on the type of economic and social structure.(1)

Even when science and technology make it possible to make use of certain resources, their utilisation may come up against barriers arising from social and cultural traditions. Thus certain types of food resource have been rejected in famine-ridden developing countries ; similarly in industrialised countries certain kinds of fish, although edible and in plentiful supply, are not consumed because they are alien to the population's eating habits. Another illustration of the importance of individual behaviour is afforded by urban refuse, which should be recycled at much lower cost if pre-sorted by households before being collected by public-disposal services.

Taxation policy clearly plays an important role by affecting the price of resources. It is sufficient here to consider the percentage levied by governments in consumer as well as producer countries in the price of petroleum products. The tax treatment accorded to oil companies by their countries of origin has moreover largely contributed to the development of their activities and has influenced their management of oil resources.

Another factor of capital importance is the legal and regulatory framework, which has proved to be one of the most effective ways of developing less polluting technologies. Other examples are the influence of speed limits on the amount of energy consumed by transport systems or the effects of building standards on thermal insulation and hence on the consumption of energy for space heating. A significant point is that in many countries the thermal insulation of buildings was the responsibility of the owner whereas the energy for heating the buildings was paid for by the tenant ; clearly providing no incentive to save energy.

The prevailing industrial structure also has a very marked effect on the management of resources. In most countries, the activities of electricity generation and distribution utilities have commonly been restricted to that energy carrier alone, often by law ; these utilities consequently had little incentive to recover and sell the residual heat of power stations, which would have meant their becoming producers and distributors of both heat and electricity. Innovations leading to improved management of electricity generating and distributing systems have moreover developed more freely in countries where these utilities enjoy a monopoly than in countries having a large number of competing firms. Similarly, the economic optimum sought by industry,

1) Cf. Part One of this volume : "The Social Function of Science and Technology".

hence the management of the resource in question, will appreciably vary according to whether the different stages of the life-cycle all take place in one sector with a high degree of vertical integration, or are broken down among a large number of firms.

All in all, the most important factor is perhaps the general economic framework in which the "rules of the game" governing the management of resources are laid down. Market mechanisms have so far been the principal frame of reference, although rare indeed have been cases where a real "market" existed in the economic meaning of the term.(1) Moreover the very nature of market mechanisms limits the extent of their application to resource management. Thus they cannot be applied to the environment, which must then be treated as external to the market. Also, if resources are to be considered as a common heritage which should then so be managed to meet the needs of all humanity and future generations, the market alone is unable to exercise what Bertrand de Jouvenel has called a tutelary function with regard to this heritage, especially as it is basically governed by short- and medium-term rather than long-term forces. It is precisely because of these market-mechanism limitations that government authorities in all countries are increasingly having to correct or supplement them in various ways (by regulatory, tax, financial or other measures) which, although their influence is increasingly felt, do not yet seem to be based on any overall view of natural resources.

D. THE INTERNATIONAL DIMENSION OF THE NATURAL RESOURCES SYSTEM

Before speaking of the international dimension of natural resources, the global dimension of the problem should first be pointed out. This global concept stands little chance of having any practical and immediate applications for resource management as long as responsibility for the latter is not entrusted to a world authority or until the physical limit of the earth's resources is about to be reached. The fact that this global dimension is not taken into account at the economic and political level means that the management of resources is strongly influenced by the distribution of power between countries. In the case of energy, for example, the greater weight of the consumer countries up to 1973 was a determining factor in the increasing exploi-

1) For a fair number of mineral resources, price and tr-de levels have been determined according to the division of power among a small number of allied transactors (an oligopolistic situation has prevailed). In a different context, reference may be made to the organisational problems of agricultural markets.

tation of the Middle East oil deposits ; it is reversal of this balance in favour of the producing countries which led to attempts at energy conservation. While this trend makes for better resource management, it is questionable whether the conditions under which it took place are the most satisfactory, either for the consumer countries, or for international economic and monetary equilibrium.

At the present time, most resource management decisions are taken in a national framework and differences between countries are substantial. While the uneven geographical distribution of all primary resources is too well known for any further comment to be needed, the demand for them and the uses to which they are put are equally diverse. This is due in particular to differences in economic development, social and industrial structures, modes of life, climate and geography.

An immediate corollary of the specific nature of national conditions is the basically international character of resources on several counts. To begin with, it will be remembered that such questions as those of marine resources or transfrontier pollution are international by definition. A second aspect is the dominant role played by multinational firms in recovering,processing and marketing a large number of natural resources, which is enhanced by their often high degree of vertical or horizontal integration, according to whether they cover all stages in the life-cycle of one resource (like the oil companies) or cover several resources (for example the non-ferrous metals sector). Another aspect of this international dimension is competition ; whether between industrial firms in different countries, between countries seeking access to certain resources, in gaining control over resource technologies between countries and in acquiring economic and social benefits from their exploitation.

The essential point is the interdependence of all countries, since no nation can hope to find all the resources it needs on its own territory. A country's inability to be entirely self-sufficient can only become more acute as its economic and social development proceeds, precisely because of the increasing variety and amounts of resources the country then needs.

This interdependence, and the disadvantages of undue dependence on some specific resource of a particular country (or group of countries) was clearly demonstrated by the oil crisis. As a result, most of the oil-consuming countries are doing their utmost to reduce their dependence on oil in general and on Middle East oil in particular. However successful these efforts may be, they do not alter the fact that the growing interdependence of all countries with respect to resources as a whole is an irreversible phenomenon, since otherwise a general recession if not utter economic breakdown may well result.

Growing interdependence in regard to supplies also implies interdependence in the use of resources, since the need to maintain the balance of trade means that a rise in imports is followed by measures to increase exports. To conclude, interdependence is not only a problem of obtaining resources and of exporting goods made from them but is also an essential feature of science and technology : quite apart from the fact that science, as knowledge, knows no frontiers, no one country possesses the scientific and technological "resources" enabling it alone to undertake all the research, and even more the development, which it needs.

o

o o

Such factors as the many and complex relationships between the different types of natural resources ; the concept of resource life-cycles, the overlapping scientific, technological, economic, social and political aspects of the resources system ; the international dimension of this system and in particular the interdependence of all countries all gain in importance as the number and volume of resources produced, traded and consumed increases i.e. as a function of economic and demographic growth. The obvious conclusion is that an overall natural resources policy has become a prerequisite for any significant improvement in resource management.

No country has yet formulated any such overall policy. At most a number of sectoral policies can be found, often in response to some crisis or serious difficulty that has arisen. Although all Member countries have long had agricultural policies (1), no real food policy yet exists. It was the growing public awareness of pollution problems which induced governments to formulate environmental policies, and the 1973 oil crisis which triggered current efforts to develop effective energy policies. Although no material resources policy yet exists, it may be noted that several Member governments are trying gradually to move in that direction. There is little if any co-ordination between these sectoral policies and between the institutions responsible for them, however, and one fundamental task of governments in coming years should be to integrate them in the wider framework of an overall natural resources policy.

Our purpose is not to analyse what these policies might be, but to show how science and technology can contribute to their formulation and implementation and to identify the main problems such action poses for science and technology policy. The following chapters are

---

1) It may be noted in this context that these policies have gained in importance as a result of economic and social problems raised by the development of agriculture.

devoted primarily to this question. It is however pointed out that the fundamental problem in any resources policy will be how to reconcile two apparently divergent but equally essential requirements : to acquire an overall view of the entire resources system while taking full account of the specific nature of each type of resource.

II

SCIENCE, TECHNOLOGY AND NATURAL RESOURCES POLICY

The contribution science and technology can make to a natural resources policy is twofold. First of all, they can help in the formulation of such a policy by promoting a better understanding of the system in all its complexity, by clarifying the nature of the problems and by showing all the interactions involved. Secondly, a traditional role they can play is to serve in implementing resources policy, as instruments for attaining specific objectives or exploring and keeping open alternative options.

These are separate but closely complementary roles. It is through improved understanding of the resources system that it will be possible to define specific R&D objectives or to identify the main alternatives to be explored. Conversely, knowledge of a particular resource or technology is necessary for an understanding of the system as a whole, and the everchanging pattern of science and technology means that our analysis of the system has to be continuously revised. This is another example of the general problem stated at the end of the previous chapter, i.e., the dual requirement of acquiring an overall view of the system while taking account of the specific nature of each resource.

A. SCIENCE, TECHNOLOGY AND THE FORMULATION OF NATURAL RESOURCES POLICY

1. Studying and understanding the resources system

The first problem in formulating an overall natural resources policy is that knowledge of the overall resources system in all its complexity is extremely limited because remarkably little research has so far been devoted to it. Improved knowledge of the system should therefore be a priority objective.

At the same time, it must be recognised that in developing research aimed at a better understanding of the system, the methodological as well as practical difficulties are considerable. As already stressed in the Brooks Report (1), "we lack an adequate science for the management and measurement of complex interacting systems".

---

1) Science, Growth and Society : A New Perspective. Report of the Secretary-General's ad hoc Group on New Concepts of Science Policy, OECD, Paris, 1971, p. 57.

From this standpoint, while the techniques of systems analysis would no doubt be very useful, the degree to which they can be utilised is open to question : to what extent can these techniques take account of political, economic and social factors which, although of fundamental importance, cannot be quantified or expressed mathematically ? There are limits to the forced application of conceptual tools borrowed from some scientific discipline, even one as universal as mathematics. An objective assessment of the various techniques of systems analysis and of their real potential for studying the resources system, therefore, seems all the more urgently needed.

Quite apart from the merits and drawbacks of particular techniques, recourse to what is conventionally known as the scientific method ought rather to be based on the fundamental principles of that method which roughly consist in repeating the following procedure as many times as necessary : the overall problem is broken down into soluble parts, the separate analyses are combined and the results compared with the observed situation, then the initial premises are modified in the light of the comparison.

In dealing with the resources system, one possible approach might thus begin with a study of the life-cycle of the different resources from the scientific, technological, economic and social standpoints, and in the light of relationships between the various resources. Another approach might be to study the implications of the various possible ways in which resources can be made to serve different functions (housing, clothing, transport, etc.). These are but two examples, but it is important to recognise that the various possible approaches are in no way contradictory. On the contrary, it is by using a number of different approaches and comparing their results that the system can be better understood.

In any event, it must be realised that the resources system is at present no better understood than were physics and chemistry in the 18th century. Any approach is therefore bound to be largely empirical, one in which the qualitative rather than the quantitative aspects will often prevail. It is therefore obvious that thorough understanding of the system will not be acquired overnight, but can only be the outcome of a long and gradual process.

It is clear that research into the system as a whole does not in any way conflict with that dealing with separate resources or technologies. On the contrary, both kinds of research are closely complementary. The system cannot be investigated in abstract terms ; research must be based on a detailed knowledge of its component parts. Conversely, as understanding is acquired, any gaps or contradictions marking R&D with respect to individual technologies and resources will become clearer. In other words, it is not a matter of placing the two kinds of research in competition with one another but of bringing them together by resorting to a permanent, profitable form of cross-fertilisation.

From a practical standpoint development of this type of research gives an added dimension to the collection, processing, storage and use not only of the scientific and technical but also political, economic and social data upon which it must be based. One present difficulty is the scattered location of establishments possessing the different kinds of information systems used. No solution will however be found to this problem without a suitable conceptual framework enabling the kind of information needed to be better defined.

Along with the methodological difficulties mentioned above, the main problem encountered in research aimed at a better understanding of the natural resources system is that these resources by their very nature require a multi-disciplinary approach. The need for interdisciplinary research and the difficulty of implementing it has been a leading problem of science policy for a number of years (1), and one which is particularly acute in the case of natural resources because of the number and variety of disciplines involved. The practical difficulties of organising joint research activities by social scientists, natural scientists and engineers (2) are well known.

The second quite obvious need for an interdisciplinary approach is due to the ultimate aim of the research in question, which is not to acquire knowledge of the system for its own sake but be able to use this knowledge in formulating a resources policy. In the field of resources, just as in other spheres of government activity, the question of relations between scientists and decision-makers arises. How can the scientists present their results in a form decision-makers can use, and how can the latter state their needs in terms that research can follow up ?

2. Prospective analysis and natural resources policy

No Member country was prepared to deal with the 1973 oil crisis. The review of the prospects offered by science and technology in the

1) In particular, this problem has been a lasting concern of the Committee for Science Policy and the Committee for Scientific and Technological Policy since the Third Conference of Ministers of Science held in 1968. Cf. Problems and Prospects of Fundamental Research in Multi-Disciplinary Fields : Brain and Behaviour - Computer Science - Materials, OECD, Paris, 1972.
Cf. The Research System, Vol 1 :Germany, France, United Kingdom, OECD, 1972; Vol 2 : Belgium, Norway, the Netherlands, Sweden, Switzerland, OECD, Paris, 1973 ; Vol 3 : Canada, United States, OECD, Paris, 1974.

2) These problems are covered in greater detail in Part One: "The Social Function of Science and Technology".

energy field (1) shows that while both hold out great promise, no marked effect on the energy situation would be felt until after 1980, because the relevant R&D was so late in starting. All research takes time, but so does industrial development of the results. For instance, it takes between 7 and 10 years to build a nuclear power station ; in a country like France, some 25 years will be needed to bring half the total number of dwellings up to the new standards of thermal insulation.

In the meantime, the situation can change considerably. In the first place, research and the outcome of the application of its results are themselves conducive to change ; the economic, political and social background of resource management can also, sometimes substantially, suddenly change, as the 1973 oil crisis has clearly shown.

The "time constant" of the natural resources system is therefore such that any resources policy must contain a long-term element based upon prospective analysis, but such forecasts cannot be reduced simply to an extrapolation of past trends to infinity. It is necessary to consider a number of possible scenarios and to pick out alternative options to be followed up, not only as regards science and technology but also from the political, economic and social points of view. In other words, prospective analysis should supplement research aimed at an understanding of the resources system by introducing, to the greatest possible extent, the variables of time and future uncertainty.

While science and technology are not of course the only factors to be considered in a prospective analysis of the resource situation, they do have a special role to play as one of the prime movers of change in the system. Moreover, precisely because of the time element in research, development and the application of their results, the very process of selecting the lines which research should follow (see below) must necessarily be a forward-looking exercise.

Prospective analyses by governments in most Member countries have so far been inadequate. Predictions have been largely based on the mere extrapolation of earlier trends, and the planning horizon has rarely exceeded 5 years. The most active institutions in this field have been a few private, non-profit organisations and above all certain large industrial (particularly multinational) firms.

Strategic planning by commercial firms, however valuable, can in no way replace governmental prospective analysis : it is not the task of any company, even the largest, to cover the whole field of resources or to safeguard political, economic and social objectives in the general interest. It is for governments to develop the prospective

1) Cf. Energy R&D : Problems and Perspectives, OECD, Paris, 1975.

analysis which is at present so conspicuously lacking. This should not however be done in isolation but should draw on the strategic planning activities of commercial firms and on any forecasting exercises carried out by other economic and social transactors (for example, trade unions or trade associations).(1) In most OECD countries, there is an increasingly general desire for public participation in the formulation of government policy, (2) and prospective analysis could be an ideal forum for such participation.

Finally, the most difficult problem lies perhaps not so much in carrying out prospective analysis as in translating it into action. This is clear from the follow-up - or rather lack of it - in the light of comments appearing in the "Report of the Three Wise Men", (3) addressed to European Governments in May 1957 :

> "In these circumstances, Europe's dependence on the Middle East can only increase, and the recent Suez crisis has warned us how serious this danger can be. As oil imports rise, the temptation to exert political pressure by this means will increase also. Not only would any interruption in oil deliveries, should this happen in a few years, threaten us with economic disaster, but from a more general standpoint, any undue dependence of powerfully industrialised countries on unstable areas is known to lead to serious trouble for the whole world."

It is interesting to note that this Report considered only one alternative source of energy - nuclear energy - and made no mention of problems associated with the environment, safety and radio-active waste, which are today in the forefront of the debate concerning the development of nuclear energy. Similar limitations are inherent in all attempts to foresee future developments, not by their own nature but also because of deficiencies in our present knowledge. This, however, is not a valid excuse for failing to undertake such forecasts, and should instead even more forcefully show the need for continually revising them, taking into account previous errors, changing situations and the progress of knowledge.

---

1) For example the French Plan, which representatives from both industry and the unions help to prepare.

2) Cf. above Part One : "The Social Function of Science and Technology".

3) An Objective for EURATOM, Report by Louis Armand, Franz Etzel and Francesco Giordani, 4th May, 1957.

B. R&D AND THE IMPLEMENTATION OF A NATURAL RESOURCES POLICY

The task of contributing to the formulation of a natural resources policy is a new one for science and technology, necessitated by the complexity of the problems involved and made possible by the progress they have achieved. Even so the importance of this new task in no way detracts from that of their more traditional function, which is to help attain the objectives of such policy. In fact, several of the questions raised below, such as the definition of R&D objectives and priorities, co-ordination within the government sector, the role of industry and its relations with the government, are traditional problems for science and technology policy-makers, but the changing situation with regard to natural resources and the need for an overall policy in this area puts these problems in a new perspective.

1. The choice of R&D directions

A look at the major directions of R&D in Member countries with regard to natural resources in recent years shows the contrast between the volume of research along certain preferred lines and the limited attention paid to others whose importance has only lately been realised, notably as a result of the oil crisis.

Thus in the case of materials, most R&D work has been concerned with the development of those which could be used under extreme conditions (materials possessing the mechanical strength and resistance to heat needed in aircraft engines and for space applications), and the use of new properties (electronic, magnetic and optical phenomena in the solid state, for instance) with a view to establishing entirely new technologies and industries. Much less has been done to reduce the amount of waste in production methods, to lengthen the life of materials and products (as by improving corrosion resistance or product design) or to recycle a higher proportion of solid wastes.

This lack of proportion is even more apparent in the case of energy. Apart from hydrocarbons, the only energy source steadily investigated during the 60s was nuclear energy - partly for reasons of national defence or prestige quite unconnected with energy problems. It is interesting that in most countries nuclear research in the civil sector has levelled off or even declined in real terms since 1967, precisely because of the very low level of oil prices. In quite a number of cases, coal research centres have been regarded as survivors from another age which one dared not do away with ; if they were kept going it was by diversifying their activities. Research activities on other energy sources (solar and geothermal energy for example) were granted only minimal funds and were rarely taken seriously, even occasionally being described as "exotic". More remarkable still was the virtually complete lack of R&D to promote the more rational use of energy.

Similarly, as regards food resources, the agricultural research carried out for several decades has been aimed at improving productivity through increasingly intensive use of energy, by making greater use of mechanisation as well as of chemical fertilisers and pesticides. The "green revolution" incidentally consists purely and simply in introducing this type of agriculture into the developing countries.

The free play of market forces has often been cited as the predominant influence on the past pattern of R&D programmes. However, much energy and materials R&D derived directly from the needs of defence and space exploration programmes and was thus free of market constraints. Even allowing for these "exceptions", there has never been a true competitive market for materials, as defined in economic theory. Prices and quantities exchanged have never been fixed as a result of competitive bidding between large numbers of independent buyers and sellers, but rather have reflected the balance of power between a small number of transactors acting in collusion. In such an oligopolistic market the existence of positive or negative "rents" nullifies the free play of market forces.

However, as far as science and technology are concerned, the most important factor is that the economic considerations largely responsible for the main lines taken by R&D were based on the implicit assumption that the type and rate of economic growth would continue in the long term without any major change and would not encounter any limits. The main objective of R&D was, therefore, to increase production of the resources required to meet increasing demand and further stimulate its growth by reducing production costs or creating new possibilities which would eventually become new requirements. Concern to conserve resources was utterly lacking, and there seemed no point in exploring and developing alternative options promising no certain economic return, while the interdependence of the various resources and technologies was largely ignored by research.

The crises and difficulties referred to in the introduction have gradually revealed these shortcomings. Growing awareness of environmental problems has made the interdependence of resources an essential factor in research. Whether industry likes it or not, it is gradually accepting the objective of developing less-polluting technologies alongside that of their economic viability. While the energy crisis has strengthened awareness of the interdependence of resources, it has above all highlighted the need to tackle two previously neglected research aspects : the exploration of alternative options and the conservation of resources.

In other words, the choice of priorities made necessary by the extent of the field involved and by the vast number of possible research directions must henceforth be based on two main types of criterion, i.e. the impact of research on the totality of the resources system and its future evolution. The basic question for science and technology policy is therefore how to identify the points of leverage of the system where R&D could lead to the most effective and timely results, both with respect to improving the management of the system or in reacting to unforeseen political, economic or social difficulties events.

Under the specific heading of R&D priorities, therefore, again emerges the earlier-mentioned need for prospective analysis and a more thorough knowledge of the resource system. This latter goal, however, calls for a long-term effort, whereas deciding on research directions is a relatively urgent task. It will, therefore, be necessary to adopt a somewhat qualitative and pragmatic approach, particularly since the choice of priorities will depend, not only on what appears to be most desirable from the standpoint of the management of the resources system, but also on whatever possibilities are offered or limitations and time requirements implied by the present state of scientific and technological knowledge. For all the OECD countries the ever-present question of establishing R&D priorities and directions now arises in new terms and at present there is no universal method or precedent that everyone could be guided by. It would therefore be well if all countries were to exchange information on the advantages and difficulties of the different national approaches to this problem.

2. Co-ordination in the public sector

To say that the choice of the main directions for R&D should be based on the impact of research on the whole of the resources system and on the latter's future development, means that these directions should together make up an overall R&D resources strategy deriving from a general resources policy, the need for which has been mentioned earlier (see Chapter I). However decentralised such a policy may be and, in particular, however important the part played by industry, it is clear that the formulation of such a strategy is a leading task incumbent upon governments.

The problem, however, is that in the public sector policy responsibility for the different types of resources is, in all countries, spread over several government departments. Although the situation differs considerably from one country to another, a number of common features can be identified. Food resources as a whole have long been the responsibility of agricultural ministries, even though some foods (fishing products for example) frequently come under another government department. In the case of the environment, most countries now have a specific ministry because of the political importance that this subject has now assumed, but in most cases these ministries by no means cover all government activities in this area.

Following the 1973 oil crisis all Member governments set about developing more comprehensive and coherent energy policies by creating new energy ministries or by strengthening and extending the powers and jurisdiction of existing institutions coming under the ministries for economic affairs or industry. In the case of materials resources, however, responsibility is still widely dispersed : in this area there has of course been no crisis comparable to the oil crisis and even if such should occur, the number and diversity of materials resources are such that it would probably not have so marked an impact on the economy.

At the level of research and development, responsibilities are often even more widely dispersed inasmuch as a large number of public research bodies, often enjoying considerable autonomy, also play a considerable part in establishing the main directions for research. Examples are certain research councils, the various national geological surveys and nuclear energy commissions.

This spread of responsibility over many different ministries and public bodies covering different aspects or sectors of the resources system and whose objectives may sometimes conflict (environment and industrial development for example) can lead to inconsistencies, if not contradictions, in government policy, particularly as regards science and technology. Formulating and implementing an overall R&D strategy in the field of natural resources therefore calls for co-ordination as well as collaboration between these ministries and public bodies.

Past experience of Member countries in other fields is sufficient evidence of the difficulties of the practical implementation of a co-ordinated and co-operative approach within the government sector. The creation of "superministries" embracing the entire field of resources, apart from being an unrealistic proposition in most countries, would doubtless do no more than shift the problem without solving it. Nor is it necessarily desirable to confine the solution of questions that may be political in nature (e.g. the balance between environmental protection and industrial development) within the bureaucratic framework of a single ministry.

Needless to say the effectiveness of the various institutions and mechanisms responsible for co-ordination will largely depend on the political and administrative conditions peculiar to each country, yet a confrontation on national experience should help to identify the advantages and limitations of some particular solution. Among the many examples of past experience worth evaluating may be mentioned the "environmental impact statement" procedure instituted in the United States by the National Environmental Policy Act of 1969, (1) with particular reference to the extent to which this procedure has succeeded in ensuring that environmental policy objectives are incorporated in the research programmes of the various federal administrations. Another type of experiment worthy of examination is that of the "horizontal" ministries set up in a few countries to give political weight to the co-ordination function.

Another question is the specific role to be played by the bodies responsible for science and technology policy in formulating and implementing an overall R&D strategy in the field of natural resources. The point is that while such a strategy should be an integral part of an overall resources policy covering many other aspects besides science and technology, there are at the same time a number of reasons why it cannot be dissociated from science and technology policies : the volume of research involved in relation to the whole of the national R&D effort, the key importance of training qualified engineers and scientists, and the fact that R&D in this area will only prove fruitful if based on the general progress of science and technology over a wide spectrum of disciplines including both the social and the natural sciences.

3. The role of industry

Industry plays a predominant part in the area of natural resources, since it is industry that extracts and converts these resources into goods or services. In addition, it carries out the major part of the relevant R&D work. This, therefore, raises the general question of compatibility between the policies of industrial firms and the overall resources policies of governments, and between the research activities of industry and R&D strategies laid down by governments. The object of an industrial firm, including its research activity, is profit and economic viability, whereas a general resources policy has to take far broader considerations into account. Even taking no more than a strictly economic view, the optimum for the system as a whole will (rarely) be the sum of the economic optima for individual firms.

For instance, we have seen that although resource conservation is becoming one of the essential objectives of an overall natural resources policy, it does not necessarily constitute a "natural" objective for industry and particularly for the industries extracting these resources. Past experience shows that it is not solely a question

1) Cf. The Research System, Vol. 3, OECD, Paris, 1974.

of promoting appropriate research but of encouraging firms to undertake such research and to apply the results achieved, and for this to work the firms themselves must find it to their advantage. In this connection, financing industrial research will often be less effective than measures affecting the socio-economic framework in which the firms operate : standards, price-fixing, fiscal measures and other regulations.

Another example relates to the need to keep alternative options open, now a particularly crucial problem where alternative sources of energy are concerned. In most cases the capital cost of exploring such options, whose economic viability is uncertain, is very high, particularly since evaluating them often means going beyond the R&D stage and setting up demonstration projects on an industrial scale. As a rule, therefore, firms can hardly be expected to shoulder the whole task themselves. While public research bodies could play a significant role, here again it may be preferable to encourage firms to take on the responsibility by financial or fiscal measures, market guarantees, etc.

To sum up, R&D strategy in the field of natural resources cannot be restricted solely to research policy ; it should be a real policy for innovation covering any and all questions involved in the practical application of R&D results. To this end, it would seem necessary to accumulate more knowledge in this field. In fact, only recently have governments sought to stimulate innovation, and more recently still to channel it into specific directions. While the array of instruments used is a vast one, (1) knowledge of the real results is still very limited, in view of the time factor involved in all socio-economic phenomena.

4. The need for "centres of scientific and technological expertise"

Although where natural resources are concerned government and industry are the main actors on the science and technology stage, it is clear, as in other fields, that the influence of public opinion is steadily increasing.(2). It is in fact in regard to various aspects of the resources area that the effects of science and technology have been most clearly brought under closer public scrutiny. Notable examples are environmental questions and the composition of foodstuffs, but it is mainly in relation to nuclear energy that the question has begun to assume full significance.

---

1) See on-going work by the Committee for Scientific and Technological Policy on policies for the stimulation of industrial innovation.

2) See Part One : "The Social Function of Science and Technology".

The application of resource technologies increasingly seems to require public acceptance,(1) and this calls for a systematic effort to inform the public, more fully than has so far been the case. Yet it should also be noted that the data and information provided by industry and even by governments is often challenged, not on objective grounds but because of the very fact that they come from industry or government, which generally have a vested interest in the technologies that are in question.

While this is hardly the place to analyse the origins of this growing mistrust, the fact must be pointed out in that it indicates the need for "centres of scientific and technological expertise" possessing the appropriate scientific and technical skills and being free to express their views uninfluenced by various pressure groups and thus in a position to enjoy public confidence.

To a large extent this has been the role of the National Academy of Sciences in the United States, but this is a unique institution and even in that country there is no assurance that it alone will be able to handle the increasing requirements placed upon it. Perhaps this function could be handed over to the universities, which would appear ideally to combine the requisite skills and freedom, but the question is how far and under what conditions universities could effectively perform this function, in view of the additional load it would impose and the need for interdisciplinarity in research. The answer is unlikely to be the same in all countries, and in some cases new institutions will have to be set up.

Yet, whatever the differences between national institutions, their functions and the problems they would be called upon to deal with will be largely common to all countries. It would seem advisable and even necessary for them to develop closer co-operation to enable exchange of information and specialists, harmonization of their respective positions and, even, the formulation of a common doctrine.

o o o

In the final account, the contribution which science and technology can make in formulating and applying a natural resources policy will depend on the availability and quality of scientific and technical personnel (scientists, engineers and technicians) in a wide spectrum of disciplines covering the social, natural, and engineering sciences. There would appear, therefore, to be an urgent need to assess future requirements in quantitative and qualitative terms and to take steps to ensure that they are met by the higher education institutions.

1) Some technologies even depend on public involvement, e.g. conserving resources and recycling waste.

III

INTERNATIONAL RELATIONS

To an even greater extent than the questions mentioned above, the question of international scientific and technological relations is a long-standing problem of science and technology policy.(1) It assumes, however, especial importance owing to the interdependence of all countries (see Chapter I), and in the light of the changes that have followed the oil crisis of 1973 is posed in new terms, particularly as political factors are closely intermingled with purely scientific and technological considerations.

The 1973 crisis marks a turning point not only on account of the already substantial effects it has had for the energy situation, but also because of the likely consequences for many other natural resources, especially mineral resources. The balance of power between countries and the pattern of international trade are now undergoing a radical and probably continuing change. While so far it is impossible to foresee just what the result of this trend will be, quite clearly the role of science and above all technology in international relations will be much more significant than in the past.

For the industrialised countries, i.e. essentially the OECD Member countries, science and technology are one of the main instruments for diversifying their sources of energy and raw materials, and hence for ultimately restoring their economic equilibrium, but this implies a vast R&D effort exceeding the capacity of each country acting by itself. In the relations between industrialised countries and the resource-producing developing countries (particularly oil), the scientific and technological potential of the former can be regarded as their principal advantage.

A. RELATIONS BETWEEN OECD COUNTRIES

While co-operation among Member countries is one of the OECD's fundamental objectives, the limitations and even the occasional failures of the past show that international scientific and technological

1) In this regard it may be noted that the subject has appeared on the agenda of all Meetings of Science Ministers at the OECD since 1963.

co-operation cannot be successful unless based on a common political will among the countries concerned rather than merely on the internal logic of science and technology, or on that of institutions set up for the purpose. In particular, even if a common feature of the OECD countries is their high rate of scientific, technological and industrial development and they constitute a relatively homogeneous group from the standpoint of economic structures and mechanisms, in the matter of natural resources their situation varies widely, with the result that their economic and political interests may appreciably differ and even diverge.

This diversity is clearly evident in the case of energy resources. Some countries, like Australia, Canada, Norway, the United Kingdom and the United States, though most are now energy importers, have a potential of varied energy resources which in the long run may enable them to become nearly self-sufficient and even exporters of energy. Their interests and objectives differ therefore from those of OECD countries which are relatively lacking in such resources. A similar situation exists with regard to materials resources, although as a whole the Member countries are consumers rather than producers.

To the difference in interests arising from the possession of resources must be added other differences resulting from the consequences of scientific and technological advances on the relative economic competitivity of Member countries.

In the face of such differences which cannot be ignored, many factors point to the need for increased co-operation in science and technology. Predominant among these factors is the general benefit to be gained by international co-operation in all areas of R&D : a greater fund of information, knowledge and know-how needed for any R&D activity can be made available ; the whole process from the research stage to practical application of the results can be speeded up ; the overall cost of R&D programmes, by using financial resources, equipment and skilled personnel more efficiently, can be reduced ; and a wider range of approaches can be tried.

These general advantages of international co-operation are of especial importance in the case of natural resources owing to the vastness of the field, the many possible directions of research and the vital need for substantially greater R&D efforts. No country alone can conduct all the research, development and demonstration programmes it needs. In particular, investment, market guarantees and other measures required for developing resources technology, (1) in the aggregate, exceed the capacity of any individual country.

---

1) See Chapter II.

Moreover, going beyond the short-term diverging economic interests of OECD countries, it should be realised that science and technology have a unique contribution to make for achieving the common goal of improved management of the overall resources system. The essentially international character of the system can never be stressed enough, and what is at stake is not merely the economic situation of countries but above all the survival and long-term adaptation of an economic, social and political system and type of civilisation common to all OECD countries.

Clearly it cannot be the task of this paper to propose detailed specific subjects for R&D on which co-operation between the OECD countries would be both desirable and feasible, nor how such co-operation should be organised. For this purpose the approach undertaken with regard to energy (1) would have to be adopted and extended to other natural resources. The main questions which arise in organising such co-operation will here merely be mentioned. In this respect, a distinction should be made between scientific and technological research, since the practical problems of each differ appreciably where international relations are concerned.

In the matter of scientific research, relations between the OECD countries are firmly established, whether through the medium of international organisations, bilateral agreements or more frequently through direct contacts between individual scientists and between research institutions. Although this form of co-operation unquestionably can and should be intensified, the problems involved are familiar and have been sufficiently analysed to obviate any further comment in these pages.

Quite apart from the research subjects on which co-ordination or co-operation among the OECD countries might prove both desirable and feasible, two essential tasks will not adequately be performed unless national efforts are confronted and integrated in an international framework, i.e. studies aiming at a better understanding of the resources system and the prospective analysis described in Chapter II. The international dimension of the resources system and in particular the interdependence of all countries are such that any analysis of the system or study of future trends would be meaningless unless undertaken in a context reaching far beyond national borders.

Unlike the case for scientific research, international co-operation in the field of technological research appears to have made little headway. Difficulties are of course increased by the fact that this type of research has the most marked short-and medium-term effects on the industrial competitiveness of countries, while it is in techno-

---

1) See Energy Research and Development : Problems and Perspectives, op. cit. , Part III, Chapter III.

logical research that the resources needed to undertake it are frequently beyond the possibilities of individual countries. It may therefore be well to review in broad terms how international co-operation in technology might be approached.

Ideally, such co-operation should take place within the framework of a resources technology policy common to the countries involved, which would, itself, stem from a common policy for natural resources.(1) Obviously, the differences in the situations and interests of Member countries make such an approach unrealistic for the OECD area as a whole. Nevertheless, in spite of the political as well as the practical problems involved, it merits serious consideration in the regional context of a group of countries whose situations and interests, if not identical, are at least complementary (e.g. the European Member countries).

As opposed to this highly integrated approach, where the central role would lie with governments, co-operation on an à la carte basis is another alternative. Here, the principal proponents would be industrial firms, who would decide whether or not to undertake co-operative research in technology according to their particular industrial and commercial capacities and interests, even though governments could exercise a significant influence in shaping the general political context of inter-firm relationships or, more actively in encouraging co-operative research activities. One advantage of this approach is its extremely pragmatic character, and it is, in fact, mainly on these lines that technological relations between OECD countries have so far developed. A serious drawback, however, is that international co-operation in technology could not then be developed within the framework of an overall view of the resources system. Moreover, such an approach might not be compatible with the present tendency towards increased government intervention in most Member countries.

The necessary development of co-operation in technology between Member countries should therefore take a middle course which would avoid the major difficulties of the two extreme approaches discussed above, whilst retaining their main advantages. It is, of course, not possible to indicate here the precise content and modalities of such an intermediate solution, since these would only emerge as a result of discussions between governments covering a wide range of issues which have not yet really been examined and which are highlighted by the deliberately oversimplified description of the two extreme approaches mentioned above.

1) Such a common resources-technology policy would, of course, need to be established in close liaison with the overall science and technology policies of the countries concerned.

The key question on which all the others depend is the extent to which the natural resources policies of countries should be compatible or converging in order to foster the common political will necessary for the development of collaboration going beyond co-operative activities on an à la carte basis. This question is fundamental in determining the countries amongst which this collaboration could be developed (all OECD countries ? a more restricted group ? bilateral agreements ?), as well as the sectors for which it might be envisaged (the entire field of resources ? a specific type of resource such as energy or the environment ? or a more limited sector ?).

The second fundamental question is the sharing of the risks and benefits of technological research programmes. To the uncertainty inherent in any research activity must be added the fact that such programmes will involve not only the setting up of pilot plants but of large-scale demonstration projects, and will call for investment, market guarantees or other measures with considerable economic implications and risks as great as the benefits which may accrue in the event of success.

In addition, past experience of co-operative industrial research, at both national and international level, has demonstrated that the firms and countries with the strongest scientific and technological potential are those likely to reap the greatest benefits from co-operative research. This is a serious problem in view of the great differences in the scientific and technological potentials of the OECD countries.

Finally, whilst governments have a leading role to play, the development and application of resources technology is principally carried out by industry. In addition to the problem mentioned in the previous paragraph (and which embraces the question of patents and licences), this raises the question of the respective roles and relations of government and industry. In this context, the differences between Member countries are considerable and contribute to the difficulty of establishing the principles and modalities on which co-operation involving industrial firms as well as governments can be based. Moreover, the fact that multinational firms play a dominant role in the field of natural resources adds another dimension to the complexity of the issue.

The importance and complexity of the questions discussed above, and the lack of knowledge concerning their multiple aspects, indicate the urgent need for further research on the subject. Such research should lead to a clearer definition of the factors that make for, or hamper, the development of international co-operation in resources technology, whether carried out within the framework of integrated policies, in the form of specific activities or of any other intermediate solution. It cannot however be too strongly emphasized that co-operation, regardless of the form it takes, will primarily depend on a common political will, based upon the recognition of a common stake in the long-term, and strong enough to surmount the difficulties due to differences in short-term situations and interests.

B. RELATIONS WITH DEVELOPING COUNTRIES

Even more than in the case of co-operation between OCED countries, scientific and technological relations between industrialised countries and developing countries cannot be considered separately from overall political, industrial and trade relations. It is accordingly difficult here to give more than a very general account, particularly as developing countries have even more disparate features than the OECD group. It is also clear that this question cannot be properly dealt with, let alone resolved, except in a forum which includes all the countries concerned.

The direct consequences of the events of 1973 affect only a small number of resources : oil and, to a lesser degree, a few mineral resources (such as phosphates and paper pulp). Yet it must be borne in mind that the situation and distribution of power are still in flux, as evidenced by the fall in raw materials prices during 1974 and the decline, however limited, in oil prices and imports at the beginning of 1975. These events do in fact mark a turning point, partly because of oil's unique importance to the economies and daily lives of all countries, but their significance must primarily be appreciated in terms of the radical changes in the overall pattern of relations between the industrialised countries and the developing countries, or at any rate with those possessing substantial resources of energy and materials.

The fundamental point is that the Third-World countries clearly want to exercise more control over the exploitation of their own natural resources and to take an appreciably larger proportion of the profits which accrue. The mining and extractive industries were first nationalised well before 1973, a process which now seems irreversible regardless of the detailed circumstances in each case. While the more recent idea of a redeployment of certain manufacturing industries closer to the sources of the energy or materials they use has yet to be given any wide practical effect, the chances are that this will be the trend during the next few decades.

This tendency for producer countries to exercise more control over a larger share of resource development and ensuing profits may often be associated with various political or ideological goals, but the underlying reason which gives it lasting substance is that the countries concerned want to take the utmost advantage of their national resources in order to initiate and sustain a development process which political and economic conditions during the last two decades have rather hindered than helped.

This trend is obviously bound to generate grave problems for the industrialised countries, by compelling them to make readjustments made all the more difficult as it usually proceeds by fits and starts. Tensions are thus unavoidable. Yet it must not be forgotten that, apart from any humanitarian considerations, in the medium and long term the economic development of the Third-World countries is a positive and even necessary factor from the standpoint of the industrialised economies, whose domestic markets for consumer goods are approaching saturation.(1) Moreover, at a time when increasing attention is being paid to environmental considerations, the assimilative capacity and existing pollution in certain industrialised countries may well impose constraints on the development of their basic industries. Transferring these industries to the developing countries could therefore be advantageous for all countries concerned, so long as this does not result in the export of pollution to the Third World.

Science and technology will become increasingly important in this ongoing process of repatterning relations between industrialised and developing countries. The Third-World countries, whether to tighten control over the exploitation of their own resources or, more generally, to embark on economic development, cannot do without the technologies evolved by the industrialised countries, even if they do manage to acquire the scientific and technological capability needed for adapting these technologies to their own requirements and conditions.

Conversely, there is no doubt that the principal "resource" of industrialised countries is science and technology, and to an even greater extent their capacity for discovery, innovation and industrial application. The last two years have therefore witnessed the development of a new style of international relations, in which the industrialised countries, in exchange for security of supply, massively contribute to the industrialisation process of certain developing countries rich in resources, thus maintaining their own level of economic activity and reducing their balance-of-payment deficits. While it is of course with the oil-producing countries that this new type of relationship has so far been established, it is likely to become general.

A number of questions are raised by the growth of such technological relations, which are outside the scope of this paper. In partic-

1) Which does not mean that there are no "pockets" of poverty and that all needs, particularly for services, are being met.

ular, this is not the place to argue whether the OECD countries should co-ordinate their action in this respect to a greater extent than hitherto. Nor can there be any valid discussion as to the form and content of these links without the participation of all the countries concerned. It must however be stressed that this type of co-operation cannot lead to any lasting equilibrium (a need shared by industrialised and developing countries alike) unless a sufficiently broad-based, long-term view is taken of relations between the two groups of countries.

Whatever form co-operation may take, it can only be fruitful if based upon a mutual recognition of the interests and objectives of all parties concerned. As we have said, the fundamental goal of Third-World countries is their economic development, and the experience of recent years clearly shows that this goal cannot be achieved by merely transferring the technologies as developed by the industrial countries to meet their own specific needs and conditions.

In this connection, the choice of technologies to be transferred in order to meet conditions prevailing in the developing countries and their adaptation to these conditions are of fundamental importance. For example, it may be asked whether nuclear energy would suit the purposes of most developing countries, the problem not so much being the enormous investment requirements of the nuclear industry as the need for specially skilled manpower and a mastery of new, complex problems which the industrialised countries themselves still do not wholly possess. More generally, the technologies developed by industrialised countries are strongly capital-intensive, while the most pressing problem for many Third-World countries is to find jobs for their plentiful unskilled manpower.

The process of industrialisation goes far beyond the introduction of technologies, and calls among other things for manpower training facilities, the development of an administrative and commercial framework and the provision of other services, without which the technologies concerned would be no more use than automobiles in a country lacking roads and filling stations. It is important to take full account of these constraints and of the unavoidable length of the whole process.

There is also the danger that the developing countries may attach undue importance to industrialisation at the expense of developing agriculture, whereas the most acute problem in many of these countries is how to feed the population. Hence they increasingly depend on certain industrialised countries which also happen to be exporters of basic foodstuffs. In many cases therefore Third-World countries should give higher preference to agricultural technologies in their relations with these industrialised countries.

Both the developing and industrialised nations will therefore have to take full account of all these questions in their international relations, otherwise the newly emergent technological co-operation may well, sooner or later, end in failure, and the greater the disappointment the stronger the feelings of resentment and tensions which will ensue. Co-operation should moreover not be restricted to resource-producing countries, but must be extended to those countries which have no abundant, readily available resources.

In the present shifting pattern of relations between developing and industrialised countries, the latter are of course strongly inclined to use (or abuse) their technological advantage to maximum effect, but in considering the options now open, it must be realised that our civilisation itself is at stake. As already pointed out (Chapter I), the interdependence of all countries in the world is inevitable, if a recession or even general economic disaster, which no country would be able to escape, is to be avoided. Depending on present decisions, interdependence will either take the form of cut-throat competition, with its concomitant risks for world peace, or of international co-operation based upon more equitable principles leading to the development of all areas in the world.

CONCLUSIONS

The various crises and difficulties to which the different kinds of natural resources have been subjected, pose serious, immediate problems calling for urgent action by governments. While science and technology can supply some of the answers, it must be admitted that in the very short term they have no ready-made solution to offer. The main reason is that R&D in the past was not designed to provide a quick reaction capability in the event of crisis. Even, if it had been, any large-scale industrial application of R&D results takes so long to achieve that its effects are felt only in the medium and long term.

Beyond these immediate problems, the most important lesson to be learned from these crises and difficulties is that a new vision of natural resources has now become necessary. In particular, they must no longer be considered as an inexhaustible store of wealth to be squandered, but as a limited heritage which must be managed to best advantage. The limits here are first of all those imposed by our management rather than the physical limits of the earth's resources.

The second basic concept in this new vision is that natural resources and methods of recovery and use are so closely interrelated as to form a complex system with many dimensions (technological, economic, social and political). The interdependence of all countries is another basic feature of the resources system.

As a result, the management of natural resources can be improved only through overall policies dealing with the system as a whole. It is within such a policy framework that R&D can be of significant help, first by clarifying the options for change, either as insurance against future crises or difficulties or in a more positive sense so as to bring about calculated improvements in resource management.

The role of R&D in implementing an overall resources policy however poses a number of problems to those responsible for science and technological policies. In the first place, the field is so vast and the possible avenues of research so many that choices have to be made and priorities established. To this end, it will be necessary to identify the "points of leverage" of the system where R&D could lead to the most effective results, allowing for the time requirements of the programmes involved.

From an institutional standpoint, the fact that governmental responsibilities for the various aspects of the resources field are dispersed among many ministries and public agencies raises the question of their co-ordination for establishing overall R&D priorities. The dominant role played by industry raises the problem of the compatibility and convergence of its R&D activities and more especially innovations, with the objectives of an overall resources policy when such compatibility can no longer adequately be ensured by market mechanisms. Here the investigation of alternatives of improved economic viability, resources conservation and environmental protection take on special importance.

Growing public awareness of the environmental and safety problems generated by resource technologies is increasingly bringing out the need for a regular dialogue with the public authorities based upon systematic, objective information : independent "centres of expertise" would appear necessary for this purpose. While the universities could play an important part in this regard, their prime function is to train the scientists and technologists upon which the R&D contribution to improving resource management will depend.

One of the most important effects of the energy crisis is a new trend in international relations ; science and technology may be expected to have an increasingly important part to play. This trend is an incentive to review technological co-operation between OECD countries, taking into account not only their differing short-term economic and political interests, but also the many factors which point to the need for increased co-operation. Questions which arise are the sharing of the potential risks and benefits of technological research, and the respective roles and relations of private companies (particularly multinational corporations) and governments in the international context.

It is perhaps in relations between industrialised and developing countries that science and technology are becoming especially important. While both are an exclusive "resource" of the former, the latter cannot do without them if they are to achieve economic and social development. A new pattern of relations is thus emerging : the developed countries providing technology to Third-World countries in return for security of their natural-resource supplies. The problems here consist in adapting technologies to the special needs and circumstances of the developing countries and in seeing how they should fit into the overall process of economic development.

For science and technology policy, these questions are neither new nor specific to the natural resources field. All these issues have been treated not only in the Brooks Report (1) but also

1) _Science, Growth and Society_, _op. cit._

by the Committee for Scientific and Technological Policy and previous Committees. In addition, most of them have already appeared on the agendas of earlier meetings of Science Ministers. The fact of their steady reappearance shows that these problems are far from being mastered and call for study in even greater depth, since they have now arisen in new terms largely as a result of the crises and difficulties affecting natural resources.

The most novel factor is that science and technology cannot be regarded merely as a means of implementing an overall natural resources policy : they also have a very important role to play in its formulation, and are essential for acquiring a better knowledge and understanding of the resources system on which any resources policy must be based. They are also key factors in the prospective analysis such policy-making requires, in view of the long lead-time of T&D as well as of the industrial application of results.

In such new tasks for science and technology, methodological as well as practical problems arise, since an interdisciplinary approach is necessary and all sectors of the economy and society must contribute. Above all, they heighten the importance of the more general question of the role which science, technology and the ministers concerned should play in overall government policy-making.

Moreover, in the field of natural resources many aspects are inherently international. This international dimension must therefore be taken into account not only in research aimed at a better understanding of the resources system but also in the prospective analysis mentioned above. To this end, a co-ordination of national efforts is clearly desirable so that the results can be integrated in an international framework.

Science and technology in the field of natural resources are therefore more necessary than ever, and while the scope of their action will evidently be far broader than before, this in itself more clearly brings out their limitations. While essential in order to define the directions and methods for better management and decision-making, science and technology cannot by themselves set the course of change. This is primarily a matter of basic political choices : by countries individually of types of economic growth and social structure, and by countries collectively of the international political and economic framework needed for organising the management of the earth's resources.

Part Three

TRENDS IN THE LEVEL AND PATTERN OF R&D EXPENDITURES IN THE OECD AREA SINCE 1971

This paper deals with four questions :

1. Has there been any "levelling off" in national R&D efforts since 1971 ?
2. Has there been any change between the balance of public and private funding ?
3. Has there been any dealine in the priority awarded to R&D by governments ?
4. Have there been any marked changes in the objectives of government R&D funding ?

As the latest standard internationally comparable statistics are for 1971 the commentary is based on data amassed from various sources by the Secretariat.(1)

1. HAS THERE BEEN ANY LEVELLING OFF IN GROSS NATIONAL EXPENDITURE FOR R&D (GERD) ?

During the early 1960's there was rapid growth in the amount of resources devoted to R&D in most OECD countries and this was expected to continue. Science policy makers in many countries were setting goals for GERD of 2.5-3.0 % of GNP.

In the mid-sixties there were signs that expenditures on R&D were decelerating in several Member countries. At first this was thought to be the temporary result of major changes in the pattern of government support for R&D (e.g. the cutback in space spending in the United States) but recently there have been signs that both government and private funds are affected. A more disturbing possibility is now being discussed, that the real science potential in some OECD Member countries may be declining. Growth in the national R&D potential is virtually impossible to measure but it cannot take place unless the annual increase in GERD is sufficient to outweigh inflation, the increased sophistication of R&D equipment, the ageing of the R&D labour force plus an allowance for the cost of adjusting the pattern of R&D inputs to changes in the structure of demand from industry and government.

Some idea about recent trends in Member countries can be obtained by looking at two indicators.

- Average annual growth rates in GERD (at 1961 GDP prices)
- GERD as a % of Gross Domestic Product (GDP).

---

1) Tables and Graphs will be found in the Annex, pp. 108-118.

In the first of these two indicators the GERD data are roughly adjusted for the effects of inflation but not for the imponderables such as sophistication or ageing. It follows that the growth rate in deflated GERD must exceed a minimum level before one can say that the national R&D potential is growing. Although it is not possible to put a precise value on this minimum level it should be possible to allocate countries, in a given period, to one of the three following classes :

The Situation	The Symptoms
i) The national R&D potential is growing	Deflated GERD grows rapidly. GERD/GDP % grows.
ii) The R&D potential is cruising	GERD grows moderately. GERD/GDP does not change significantly.
iii) The R&D potential is declining	Deflated GERD grows very little or declines. GERD/GDP declines.

At the turn of the decade four of the countries responding to the OECD 1971 survey were in the third of the above situations. In Canada, France, the United States and the United Kingdom, there was little or no growth in GERD at 1961 prices and a decline in GERD as a percentage of GDP. In these countries the national science potential was in decline. In three countries, Greece, Iceland and the Netherlands the national science potentials could be said to be "cruising". In all the other countries GERD continued to grow at more than eight percent p.a. and the GDP % rose.

GERD(1) data have been available since 1971 for 16 countries which can be considered as a reasonable sample of OECD Member countries as follows :

To 1975 : Austria

To 1974 : Germany and the United States

To 1973 : Belgium, Finland, France, Iceland, Italy, Norway and Sweden.

To 1972 : Canada, Japan, the Netherlands, Portugal, Spain and the United Kingdom.

In Canada, France, the United Kingdom and the United States, national R&D potentials continued to decline. The % of GDP devoted to R&D fell and the average annual growth rates in GERD were very low or negative.

---

1) Or corresponding national equivalent. The data for later years are generally preliminary national estimates or projections. Further details are given in the Annex ; see "Sources and Methods".

In the Netherlands and Iceland, the growth rate of GERD dropped to well below 5 % p.a. and the GDP % fell which suggests that their science potentials had gone into decline. Norway joined this group in 1973 and the growth in German GERD slowed to "cruising speed" in 1973 and decelerated further in 1974.

In the eight remaining countries, GERD continued to grow at well over 5 % p.a. (over 20 % in Spain) and the GDP percentages usually also increased. However, with the exception of Italy, the growth rates for 1973/74 were lower than those for the earlier years.

Thus, by the end of the period GERD was "levelling off" or was in decline in half the countries for which data was available and though science potentials in the others were still growing there were some signs of deceleration.

2. HAS THERE BEEN ANY MAJOR CHANGE IN THE BALANCE OF PUBLIC AND PRIVATE FUNDING ?

For various reasons, the structure of funding of R&D is relatively rigid in the short term, changing very slowly in response to re-orientations of science policy, though there are often slight oscillations in the contribution of the two major sources of funds, government and industry, from year to year.

During the 1960's there was a clear trend towards increased participation by industry. Belgium and Germany were the only countries where the share of GERD financed by government grew perceptibly. The increase in business funding was particularly marked in the United States, Ireland and the industrialising Member countries. It was also significant in Denmark, France, Sweden and Switzerland but was slight in Austria, Norway, and the United Kingdom. In the remaining countries, including Japan and the Netherlands, there was little change in the balance of funding of GERD.

The most recent data, available for 16 countries, modify this picture of increasing participation by industry. In half the countries the percentage of GERD financed by government rose perceptibly after 1971. They were Canada, Finland, Iceland, Italy, Norway, Portugal and the United Kingdom and Germany (until 1973). The business enterprise contribution to GERD grew in Belgium, France, the Netherlands, Spain, Sweden and the United States and Austria. There was virtually no change in Japan.

There does not seem to be any direct link between these changes in the pattern of funding and the preceding analysis of "levelling off". In half of the "levelling off" countries the share of GERD financed by governments rose, in half it fell and the same was true of the countries where science potentials were still growing.

Recent data give little or no support for the view that industry may be taking over the burden of R&D funding from government in the "levelling off" countries. Even in the four of these countries where the business enterprise contribution to GERD increased there was at best a slight increase in the corresponding % of GDP. This occurred in Germany in 1974. There was no change in R&D funds as a % of GDP in the Netherlands and the United States, and a fall in France. This indicator also fell in the other four "levelling off" countries (Iceland, Norway, Canada and the United Kingdom). There was a clear increase in business enterprise R&D funds as a % of GDP after 1971 in five of the eight countries where R&D potentials were still rising (Austria, Belgium, Japan, Spain and Sweden).

3. HAS THERE BEEN ANY DECLINE IN THE PRIORITY AWARDED TO R&D BY GOVERNMENTS ?

We have virtually no information about industrial R&D after 1973. Some more recent data on trends in public support for R&D can be obtained from government science budgets. Data after 1973 are available for the following 12 countries :

To 1974 (or 1974/75) : Denmark, France, Germany, Ireland, Italy, the United Kingdom and the United States.

To 1975 (or 1975/76) : Austria, Belgium, Finland, the Netherlands Norway and France (partial).

One measure of the priority awarded to publicly funded R&D is its share in the total government budget. Since 1973 the percentage has declined in two-thirds of the countries ; there was no change in three (Ireland, Norway and the United States) and a slight rise in two (Austria and Finland).

Between 1973 and 1974 the increase in government R&D funding at fixed (GDP) prices varied generally between + 5 % and - 5 %. Three countries, Austria, Belgium and Finland, experienced growth of more than 5 % in their R&D budgets but in Italy and Denmark declines in excess of 5 % occurred. Norway and Finland were the only two of the seven countries for which projections for 1975 are available where any real increase for government support for R&D was expected.

Thus, at fixed prices there has probably been at best a modest increase in the real amount of resources devoted to R&D by governments since 1973. In these circumstances R&D programmes which maintain or see a decline in their share of total government R&D funds may suffer an absolute decline in their activities. This is a very different situation from the 1960's, when virtually all types of R&D programmes grew continuously, though at varying speeds.

4. HAVE THERE BEEN ANY MARKED CHANGES IN THE OBJECTIVES OF GOVERNMENT R&D FUNDING ?

At the time of the last ministerial meeting on science it was the general opinion that the 1970's would be the decade of the social objectives as far as government R&D was concerned. Most writers foresaw an increasing priority for civil and particularly for social programmes at the expense of military and big science projects which seemed, anyway, to be levelling off. Some even went as far as to predict that the social objectives would take over from nuclear, industrial and defence programmes as the biggest R&D spenders in the 1970's.

A percentage distribution of total government R&D funding by objective 1970 to 1974 or 1975 reveals how far these predictions have been fulfilled. Data are available from a report on the European Economic Community to 1974 and from national sources for Belgium to 1974, the United States to 1974/75 and for Finland, France (partial), Norway and Sweden to 1975.

We shall examine the percentages committed to R&D principally in support of defence, of the social objectives and of energy programmes. The balance of R&D funding goes mainly for economic objectives (agriculture, industry, etc.) and for the advancement of science (basic research and/or university research).

i) Trends in the Balance Between Military and Civil R&D

During the 1960's the share of all government R&D funding going to defence programmes tended to decline notably in the countries where such programmes were very large at the beginning of the period i.e. the United States, the United Kingdom, France and Sweden. This decline seems to be over. The defence percentage picked up again clearly in Sweden after 1970, in a less marked manner in the United Kingdom from 1971 and the United States in 1974/75. In France there has been no further decline in the defence percentage since 1972. The trend towards civil spending seems thus to have ended.

ii) Trends in Government Support for R&D for the Social Objectives

In the absence of a single standard definition of the "social objectives", data has been grouped for government R&D programmes with a direct social orientation, such as health, social services, education, etc. (1)

1) For details by country, see in the Annex "Sources and Methods".

Government funding of R&D for the social objectives grew rapidly during the sixties, as their very low initial level permitted spectacular growth rates. The priority these objectives received within total Government R&D funding continued to grow until 1971 in nearly all the countries for which information is available. Between 1971 and 1973 the rise in the share of the social objectives in total government support of R&D halted in two-thirds of the countries studied and actually fell in some. An increase may be noted, however, in Belgium, the United States and to a lesser degree in Norway. In 1974, the share of R&D on socially-oriented objectives increased in several countries (Denmark, Ireland, Italy and Sweden) where it had declined during the previous three years. Conversely, this share declined slightly in the United States.

Nonetheless, during the last years for which data are available, it may be said that the share of socially oriented programmes never reached a quarter of government R&D funding in Member countries, and in half of them it was less than a tenth.

iii) Energy

Historical international series on energy R&D deal exclusively with nuclear programmes. Nuclear R&D was the number one objective in the OECD countries taken together in the 1960's, though funding was levelling off at the end of the period. In only two Common Market countries, Belgium and Italy, did nuclear R&D increase its share of all government R&D funding between 1973 and 1974.

Changes in the pattern of government support of R&D resulting from the energy crisis are only just becoming visible. The first major impact of the crisis was felt after 1974 budgets had been prepared in most countries. The data at present available for 1974 has generally not been corrected for subsequent budget cuts and re-arrangement. Only in the five countries where a breakdown of 1975 budget data is available, could we expect to see any major changes.

In the case of Norway and Sweden, total energy R&D is not separately available. In Finland, energy programmes absorbed 3.7 % of all government R&D funds in 1973, falling to 3.4 % in 1975.

In the United States the percentage of all government R&D funds going to energy rose from 2.6 % in 1972/73 to 5.1 % in 1973/74. The share going to fossil, solar and geothermal energy has grown steadily. From very small beginnings the increase in the absolute amount devoted to solar and geothermal energy has been very rapid since 1972/73.

In France, the percentage of all government funded R&D expenditures subject to inter-ministerial discussion going to energy programmes grew between 1974 and 1975 from 28 % to 31 % with electro-nuclear programmes getting about 90 % of the funds. The share of electro-nuclear programmes in the R&D budget rose from 25 % to 27 %, and that of other energy programmes, from 3 % to 4 %.

In the United States the increase in energy R&D since 1973-74 seems to have been very largely at the expense of space and health programmes. In France, the share of "socio-economic" objectives did not fall significantly between 1974 and 1975, the main casualties being the national computer and electronic component programmes and to a lesser extent space and basic research programmes.

Annex

CONTENTS

A. GRAPHS AND TABLES

Graph I. GERD at 1961 (GDP) Prices in Millions $ US......... 108

Table 1. Annual Growth Rates (or average annual growth rates) of GERD, at 1961 (GDP) Prices Since 1969 in OECD Member countries 109

Graph II. GERD as a percentage of GDP.......................... 110

Graph III. GERD by Source of Funds (in %) 111

Table 3.A. GERD by Source of Funds since 1971 (in %) 112

Table 3.B. R&D Financed by Business Enterprises as a percentage of GDP Since 1971 113

Graph IV. Total Government R&D Funding in Millions $US (at 1961 prices) 114

Graph V. Total Government R&D Funding as a percentage of Total Government Expenditure 115

Table 5. Total Government R&D Funding in Selected Member Countries as a percentage of Total Government Expenditure since 1973 116

Table 6. United States - Share of Selected Functions in Total Federal Conduct of Research and Development - in % since 1972/73 117

Table 7. The French Civil "Science Budget" for 1974 and 1975. 118

B. SOURCES AND METHODS

I. Gross National Expenditure on R&D and its Distribution by Source of Funds 119

II. Total Government R&D Funding and its Distribution by Objective 124

III. Non-R&D Data 128

Graph I

GERD AT 1961 (GDP) PRICES

in millions US $

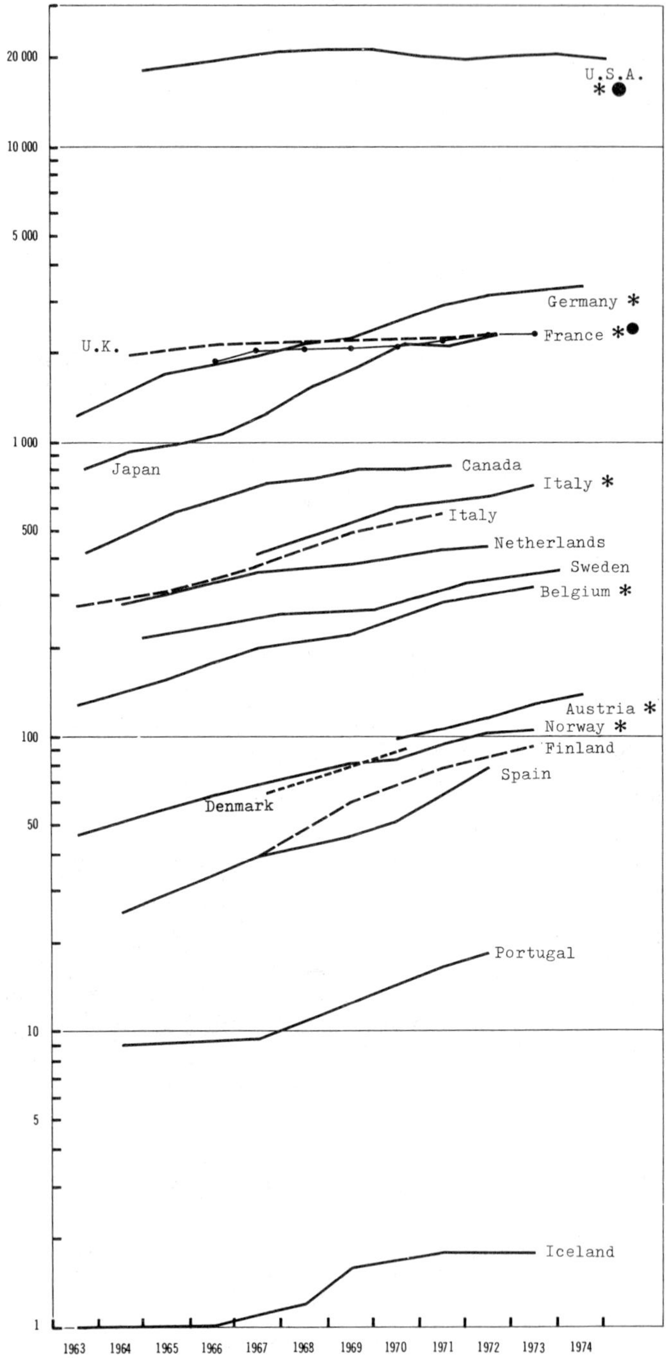

* These data include social sciences and/or humanities.

● National total approximating to GERD.

Table 1

ANNUAL GROWTH RATES (OR AVERAGE ANNUAL GROWTH RATES) OF GERD, AT 1961 (GDP) PRICES SINCE 1969 IN OECD MEMBER COUNTRIES

	1971/69 (c)	1972/71	1973/72	1974/73
Austria (a)	8.7	12.1	9.4	8.0
Belgium (a)	13.5	← 6.5	→	..
Canada	1.5	-0.8	..	..
Finland (a)	14.5	← 8.8	→	..
France (a)(b)	3.2	3.3	0.8	..
Germany (a)	11.9	7.9	5.1	-0.5
Iceland	4.6	← 1.1	→	..
Italy (a)	8.6	3.9	10.1	..
Japan	8.9	8.2	..	..
Netherlands	5.5	2.8	..	..
Norway (a)	8.4	9.7	2.8	..
Portugal	..	9.3	..	..
Spain	13.5	22.4(d)	..	..
Sweden	11.4	← 6.5	→	..
United Kingdom	← 1.5	→		..
United States (a)(b)	-3.8	2.7	1.7	-3.7

N.B. The implicit GDP deflator reflects changes in overall prices.

a) Includes social sciences and humanities (United States : social sciences and psychology).
b) National total which differs from GERD as defined for OECD surveys (See Sources and Methods).
c) Austria 1971/70 ; Spain 1970/69.
d) 1972/70.

Graph II

GERD AS A PERCENTAGE OF GDP

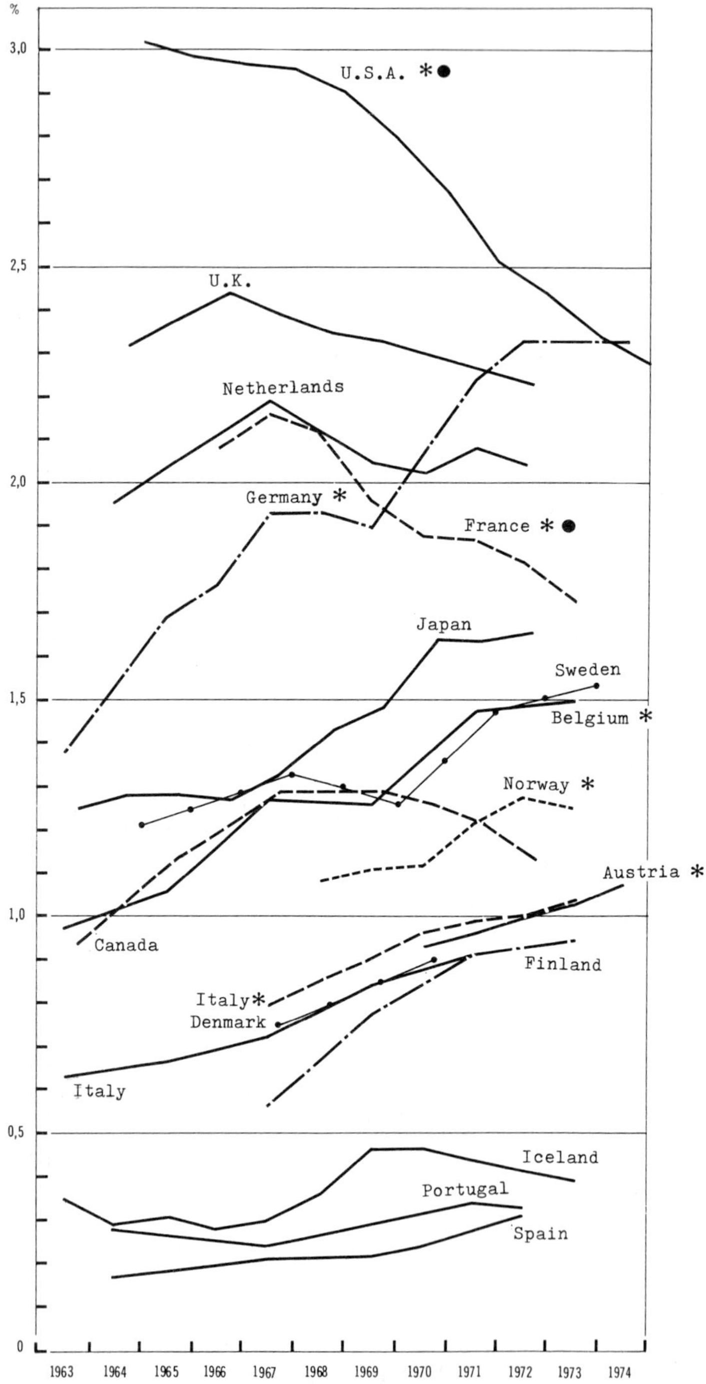

* These data include social sciences and/or humanities.
● National total approximating GERD.

Graph III

GERD BY SOURCE OF FUNDS (%)

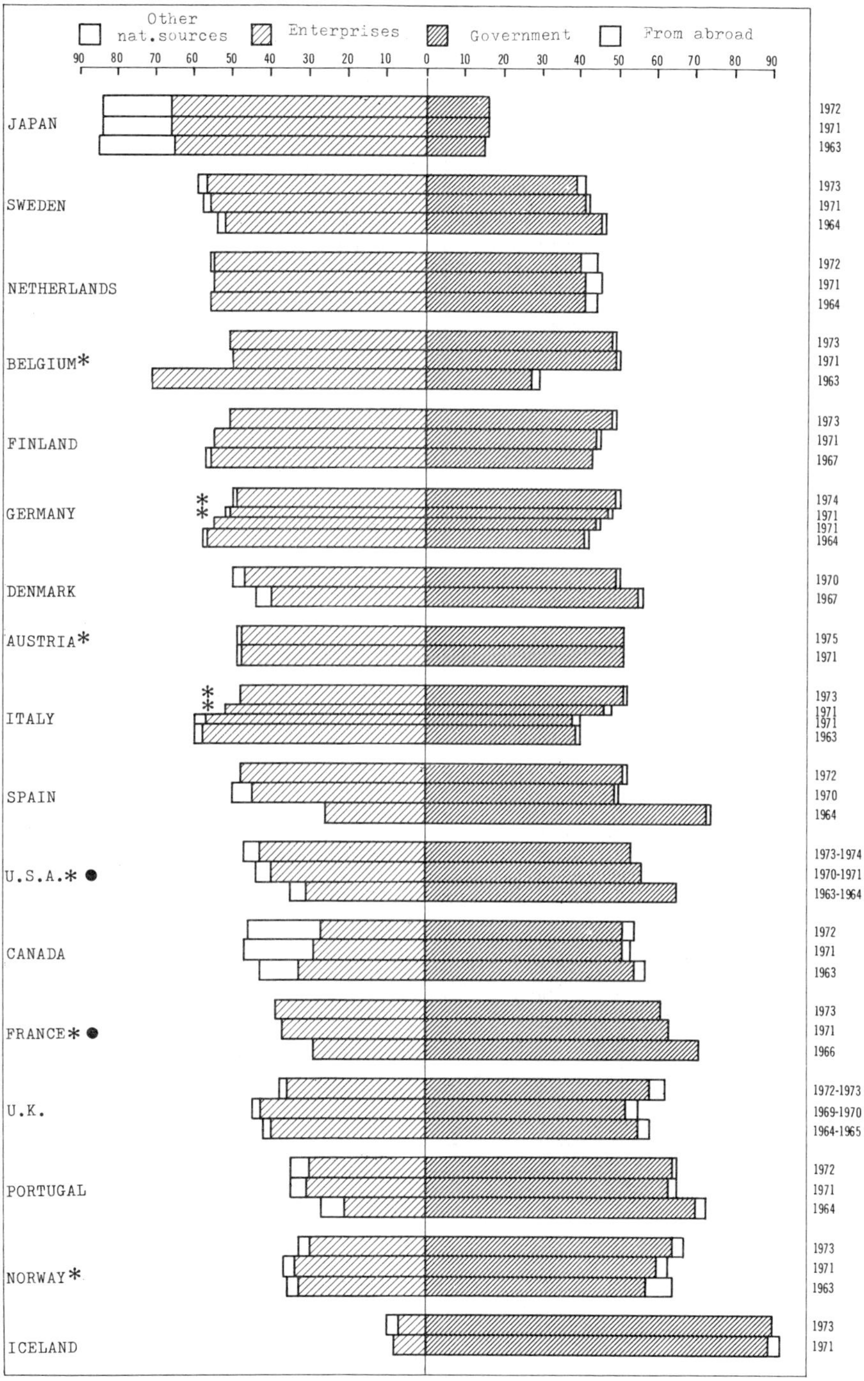

* These data include social sciences and/or humanities.
● National total approximating GERD.

Table 3.A

GERD BY SOURCE OF FUNDS SINCE 1971

Percentages

		Government	Business Enterprise	Other National	Abroad
Austria (a)	1971	50.5	48.4	1.1	
	1972	52.9	46.0	1.1	
	1973	52.5	46.5	1.0	
	1974	51.8	47.3	0.9	
	1975	51.4	47.8	0.8	
Belgium (a)	1971	48.5	50.1	. (c)	1.4
	1973	48.0	50.6	.	1.4
Canada	1971	50.6	29.3	17.7	2.4
	1972	50.8	27.4	18.6	3.2
Finland (a)	1971	44.3	54.8		0.9
	1973	48.1	51.2		0.7
France (a)(b)	1971	62.7	37.3	. (c)	..
	1972	61.9	38.1	.	..
	1973	61.4	38.6	.	..
Germany (a)	1971	47.0	50.5	1.0	1.5
	1972	48.1	49.6	1.0	1.3
	1973	48.8	49.0	1.1	1.1
	1974	48.6	49.3	1.1	1.0
Iceland	1971	89.2	8.0	-	2.9
	1973	90.0	7.0	.	3.0
Italy (a)	1971	45.9	52.2	. (d)	1.9
	1972	46.1	52.3	.	1.6
	1973	50.5	48.4	.	1.1
Japan	1971	15.7	66.4	17.8	0.1
	1972	15.8	66.4	17.7	0.1
Netherlands	1971	40.9	55.1	0.4	3.6
	1972	39.8	55.7	0.5	4.0
Norway (a)	1971	60.5	33.7	3.4	2.7
	1972	62.6	31.5	3.5	2.4
	1973	63.8	30.4	3.0	2.9
Portugal	1971	63.4	30.9	4.1(e)	1.5
	1972	64.7	30.1	4.7(e)	0.6
Spain	1970	51.1	45.5	4.7	1.0
	1972	50.8	47.7	0.4(e)	1.0
Sweden	1971	41.2	55.8	2.0	1.0
	1973	39.4	57.5	1.8	1.3
United Kingdom	1969/70	51.6	42.7	2.0	3.7
	1972/73	58.0	36.0	2.0	4.0
United States (a)(b)	1971	56.0	40.4	3.6	-
	1972	55.9	40.5	3.6	-
	1973	54.1	42.3	3.6	-
	1974	52.9	43.4	3.7	-

a) Includes social sciences and/or humanities.
b) National total differs from GERD.
c) Distributed between government and business enterprise.
d) Included in government.
e) Private Non-Profit only.

Table 3.B

R&D FINANCED BY BUSINESS ENTERPRISES
AS A PERCENTAGE OF GDP SINCE 1971

	1971	1972	1973	1974
Austria	0.46	0.46	0.47	0.51
Belgium (a)	0.72	..	0.76	..
Canada	0.36	0.31	..	..
Finland (a)	0.50	..	0.48	..
France (a)	0.70	0.69	0.67	..
Germany	1.13	1.16	1.14	1.15
Iceland	0.04	..	0.03	..
Italy	0.52	0.52	0.50	..
Japan	1.09	1.10	..	..
Netherlands	1.14	1.14	..	..
Norway	0.41	0.40	0.38	..
Portugal	0.10	0.10	..	..
Spain	0.11(b)	0.15	..	..
Sweden	0.83	..	0.89	..
United Kingdom	1.00(c)	0.80	..	..
United States	1.00	0.99	0.99	0.99

Business enterprise are considered to finance little or no R&D in the social sciences and humanities.

a) Includes a small element of other private funds.

b) 1970.

c) 1969/70.

Graph IV

TOTAL GOVERNMENT R&D FUNDING

in millions US $ at 1961 prices

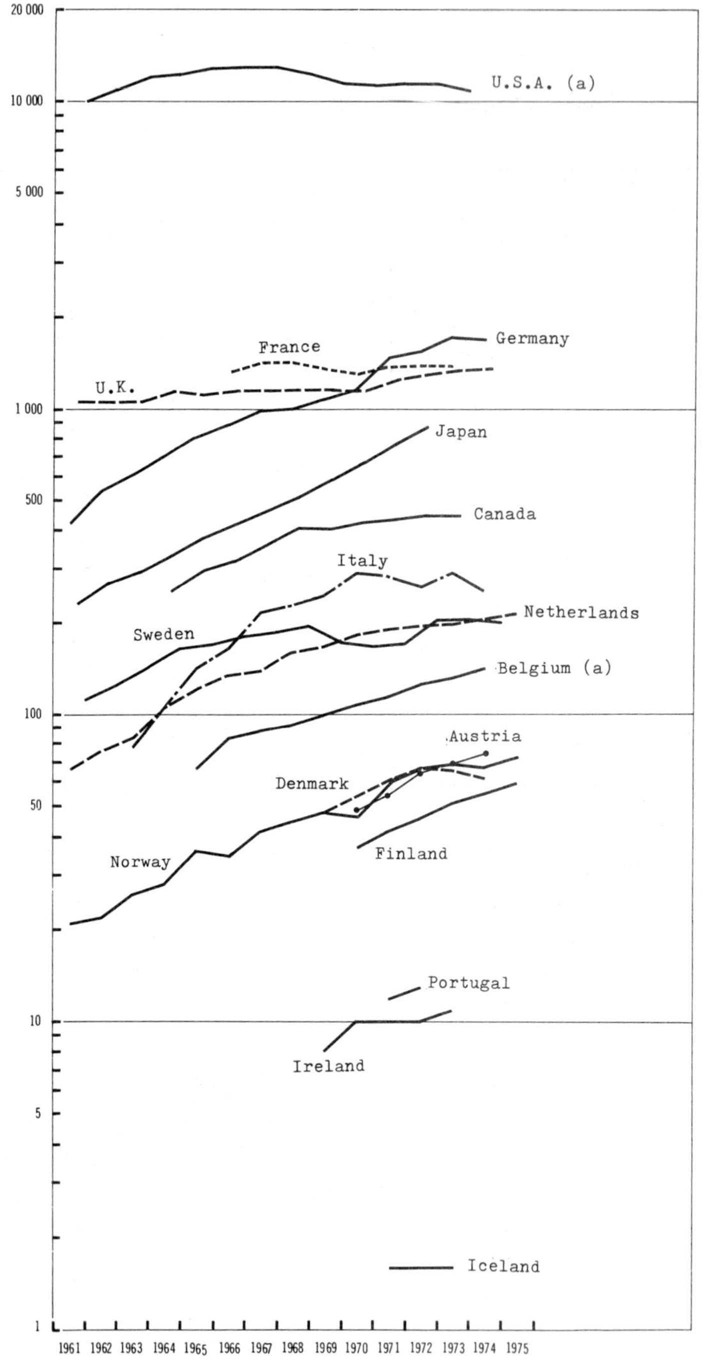

a) Current expenditure only.

Graph V

TOTAL GOVERNMENT R&D FUNDING AS A PERCENTAGE OF TOTAL GOVERNMENT EXPENDITURE

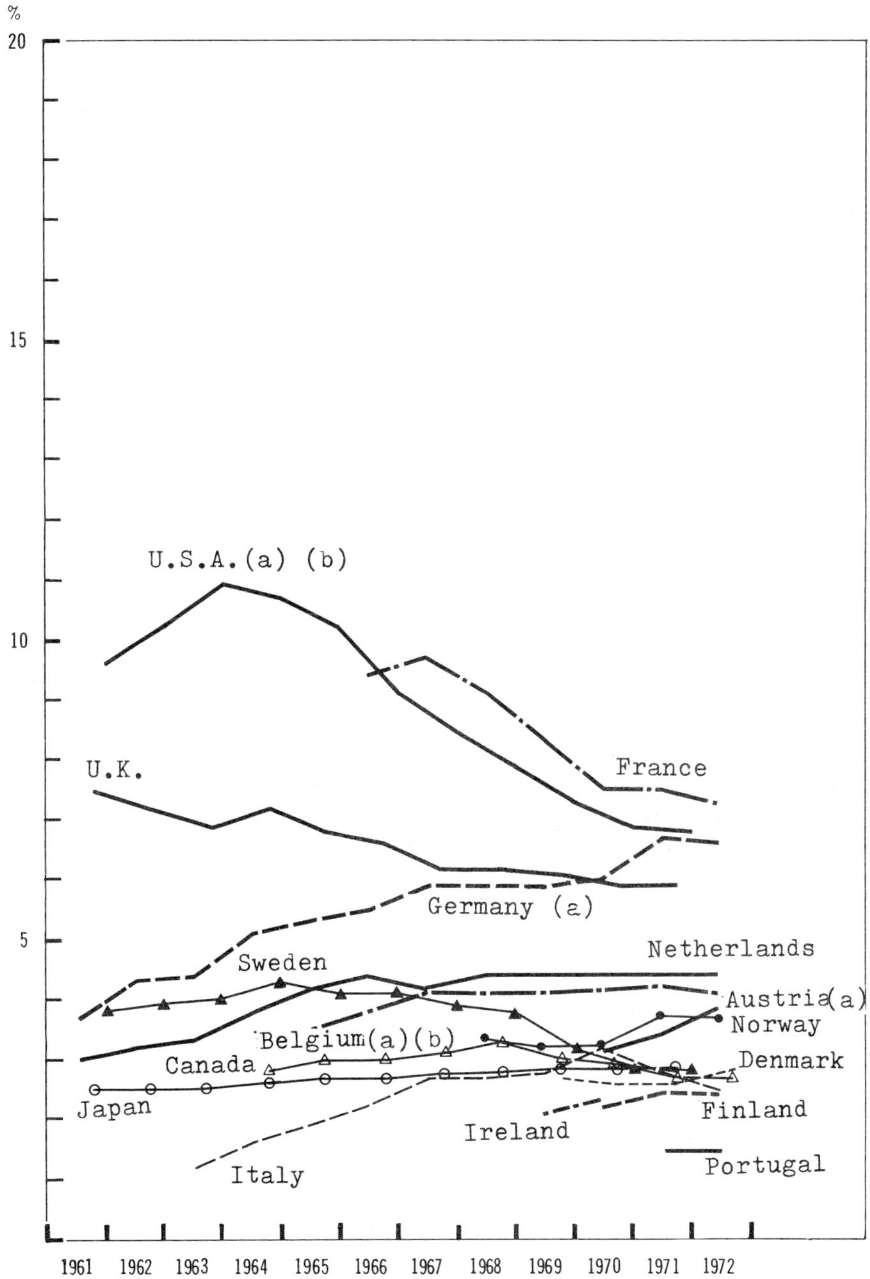

a) Government final consumption expenditure only.
b) Current expenditure only.

Table 5

TOTAL GOVERNMENT R&D FUNDING IN SELECTED MEMBER COUNTRIES (a) AS A PERCENTAGE OF TOTAL GOVERNMENT EXPENDITURE SINCE 1973

	1973 (73/74)	1974 (74/75)	1975 (75/76)
Denmark	2.0	1.8	..
Germany	5.1	4.9	..
Italy	1.5	1.4	..
Netherlands	3.5	3.4	..
United Kingdom	3.3	3.2	..
Austria	1.6	1.7	1.7
Belgium	2.6	2.6	2.5
Finland	7.9	8.7	8.3
France	5.6	5.3	..
Ireland	1.4	1.4	..
Norway	2.6	2.6	2.6
Sweden	3.1	2.8	2.8
United States	5.9	5.9	..

N.B. These data are comparable over time but generally not between countries.

a) OECD National Accounts data for total government expenditures end with 1972. The data in the upper part of the table show total government R&D funding as a % of total central government budgets as quoted in a recent EEC publication. In the lower half various national measures of total government expenditures taken from budgets have been used. (For details see Sources and Methods).

Table 6

UNITED STATES - SHARE OF SELECTED FUNCTIONS IN TOTAL FEDERAL CONDUCT OF RESEARCH AND DEVELOPMENT

Percentages

By Selected Function	1972/73	1973/74 (est.)	1974/75 (est.)	1973/74 if 72/3=100	1974/75 if 72/3=100
Energy Development and Conversion	2.6	3.2	5.1	130	226
Nuclear	2.1	2.4	2.9	117	157
Fossil	0.3	0.4	1.1	161	481
Solar	n	0.1	0.2	340	1130
Geothermal	n	n	0.1	345	1990
Other	0.2	0.3	0.7	209	532
Health	9.5	11.7	10.0	131	123
National Defense	53.5	51.7	52.1	102	114
Space	15.5	14.1	13.0	96	98
Environment	4.0	4.2	5.0	109	145
Education	1.4	1.3	1.1	98	91

Source : An Analysis of Federal Funding by Function, 1969-1975, NSF 74-313.

Table 7

THE FRENCH CIVIL "SCIENCE BUDGET" FOR 1974 AND 1975 [a]

Percentages

	1974		1975		1975 where 1974 = 100	
Energy	27.6		31.4		126.7	
Of which : Nuclear		24.7		27.5		123.6
Other		2.8		3.9		153.7
Major Programmes	22.5		20.4		100.4	
Space		14.6		13.8		105.3
Computers		7.3		6.6		100.7
Electronic Comp.		0.7		-		0
Other	49.9		48.2		107.3	
Basic research		27.7		25.7		106.7
Socio-economic		9.0		8.7		108.1
Industrial		13.3		12.9		108.1
Residual		0.9		0.8		103.6
Total	100.0		100.0		111.1	

a) The budget covers only newly-committed funds for R&D programmes subject to inter-ministerial discussion. It excludes defence, major aviation programmes, some telecommunication programmes and university R&D financed through the general grant from the Ministry of Education.

Source : "L'enveloppe-recherche en 1975", Le Progrès scientifique N° 173, novembre-décembre 1974, DGRST, Paris.

B. SOURCES AND METHODS

Sources and methods are discussed under three headings :

I. Gross national expenditure on R&D (GERD) and its distribution by source of funds.

II. Total government R&D funding and its distribution by objectives.

III. Non-R&D data.

I. GROSS NATIONAL EXPENDITURE ON R&D (GERD) ITS DISTRIBUTION BY SOURCE OF FUNDS

The standard parameter for measuring the financial resources a country devotes to R&D is "Gross National Expenditure on R&D" (GERD). GERD includes current and capital expenditures on R&D performed on national territory as reported by the performing agency. The OECD has recently changed its standards for the coverage of GERD in terms of fields of science. Previously only R&D in the natural, medical and agricultural sciences and engineering was included. In forthcoming surveys R&D in the social sciences and humanities will be included. As the purpose of this report is to examine trends rather than to compare levels of R&D expenditures in Member countries, the Secretariat has used the longest time series available for each country, which may or may not include R&D in the social sciences and humanities. In one or two cases the national total varies slightly from GERD as defined above.

Five sources of funds for GERD are normally identified : Business Enterprise, Government, Private Non-Profit, Higher Education and Abroad. In this report Private Non-Profit and Higher Education are grouped as "other national". R&D performed in universities, financed out of their general grant from the Ministry of Education, has been credited to government as a source of funds.

GERD is an aggregate of data for performing sectors. In some Member countries, surveys of the different sectors do not cover identical time periods (i.e. fiscal year for Government, calendar year for Enterprises and "academic year" for Higher Education). The period covered by GERD in these countries is shown in Tabulation A. The same percentages were used to reweight the GDP and deflator data for these countries.

A. General Remarks on Trends in the OECD Area in the 1960's

These are based on Patterns of Resources Devoted to Research and Experimental Development in the OECD Area 1963-1971 (1). In this report GERD excludes R&D in the social sciences and humanities.

B. Data Quoted in Graphs I, II, III and Tables 1 and 3

The latest international standard survey data available are for 1971. Three principal sources have been used for more recent GERD data :

i) National publications
ii) Data supplied by the national authorities for the present document
iii) Table T6 of the OECD questionnaire of the 1971 ISY.

1. Coverage

For coverage of the data used for GERD and for remarks on the breakdown by source of funds see Tabulation A.

2. Sources

Austria :

Preliminary data for the 1975 Report on Research of the Federal Government to Parliament, submitted by the national authorities for the present document and adjusted by the Secretariat to exclude Federal transfers to abroad (no data available on transfers to abroad of the Länder and other sectors).

Belgium :

Revised data 1963-1973 submitted by the national authorities for the present document.

Canada :

Data supplied by the Education, Science and Culture Division of Statistics Canada.

Finland :

Data submitted by the national authorities for the present document.

France :

Data submitted by the Délégation générale à la recherche scientifique et technique (DGRST) for the present document.

1) OECD, Paris, 1975.

Germany :

Preliminary data for the "Fifth Federal Report on Research" submitted by the national authorities for the present report and adjusted by the Secretariat to exclude public and private transfers to abroad.

Iceland :

1963-71 : National response to Table T6 of the questionnaire for the International Survey of Resources Devoted to R&D in 1971 in OECD Member Countries (1971 ISY)

1973 : National publication "Rannsóknastarfsemi á Íslandi á Svidi Raunvísinda 1972 og 1973". (National Research Council, Reykjavik, January 1975) + OECD estimate to exclude the social sciences and humanities.

Italy :

Two series of data have been used :

a) Data revised by the Italian National Institute of Statistics for 1963-1971 ;

b) The national response to Table T6 of the 1971 ISY questionnaire for 1967-70 with updated data for 1971-1973 published in "Statistiche sulla ricerca scientifica, Anni 1971-1973 (Estratto da ISTAT, Bolletina mensile di statistica n° 1, Gemaio 1975, Appendice III).

Japan :

"Report on the Survey of Research and Development in Japan, 1973", Bureau of Statistics, Office of the Prime Minister (March 1974) Summary Table 2-2.

Netherlands :

1964-1971 : Netherlands' response to 1971 ISY Table T6.

1972 : "Speur-en Ontwikkelingswerk in Nederland 1972" (Centraal Bureau voor de Statistiek, 1975).

Norway :

1963-1972 : NAVF, NLVF, NTNF : "Fou-statistikk - Utgifter og Personale". Annual reports 1969, 1970, 1971 and 1972.

1973 : Preliminary data for the 1973 Report provided for the present document.

Portugal :

1964, 1967 and 1971 : Portugal's response to 1971 ISY Table T6.

1972 : Data provided by the national authorities for the present document.

Spain :

1964-1970 : National responses to the OECD ISY's.

1972 : Data provided by the national authorities for the present document.

Tabulation A - COVERAGE OF GERD

Country	Periodicity of Data		3. Data comparabl with ISY Data (1)
	1. Years Available	2. Coverage of GERD Quoted as Year X	
Austria	Annually 1970 to 1975	Calendar x	No
Belgium	Biennially 1963-73	Calendar x	* (yes)
Canada	Annually 1963/64 - 1972/73	x 85% (2) x +1 15%	yes
Finland	Biennially 1967-1973	Calendar x	No
France	Annually 1966-1973	Calendar x	No
Germany	1964, 1967, 1969, 1971 Annually 1970-1974	Calendar x	No
Iceland	Annually 1961-1971 and 1973	Calendar x	Yes
Italy	1. Biennially 1963-1971	Calendar x	Yes
	2. Annually 1967-1973	Calendar x	No
Japan	Annually 1963/1964 1972/1973	x 75% x+1 25% (2)	Yes
Netherlands	1964, 1967 Annually 1969-1972	Calendar x	Yes
Norway	1963,1966 Annually 1968-1973	Calendar x	No
Portugal	1964, 1967, 1971, 1972	Calendar x	Yes
Spain	1964,1967,1969,1970, 1972	Calendar x	Yes
Sweden	1964, 1967, 1969, 1971 and 1973	* Varies (2)	Yes
United Kingdom	1964/65 1966/67 - 1969/ 1970 - 1972/73	*x-1 3-4 % x 89-90% x+1 6-8 % (2)	Yes
United States	1963/64 - Annually 1973/74	x-1 15 % x 85 % (2)	No*

1) Where no other explanation is given non-comparability between the present data and IS data is mainly due to the inclusion of R&D in the social sciences and humanities in t former series.

AKDOWN BY SOURCE OF FUNDS

Coverage of GERD		6. Breakdown by Sources of Funds
ocial Sciences and Humanities	5. Other Remarks	
cluded	In part new calculations which for methodological reasons, are not comparable with previously submitted ISY data	It was not possible to break down the national class "other sources" between "other national" and "abroad"
cluded	* Revised ISY data	Data for "other national" is not separately available. This expenditure is included in either "Government" or "Business Enterprise" according to whether its original source was private or public
cluded		
cluded		Data for "other national" not separately avail. "Business Enterprise" includes funds from PNP organisations & the "own funds" of Higher Ed.
cluded	Not GERD but total national expenditures excluding funds from abroad; including expends to abroad	"Abroad" is excluded. Funds from PNP & own funds of private universities are included in "Business Enterprise". Own funds of public universities are included in "Government"
) 1964-1971 Excluded) 1970-1974 Included	Coverage of the Government and PNP sectors has changed between the two series of data	"Other national" includes funds from scientific institutions outside the Higher Ed. sector which are mainly state financed
cluded		
cluded		"Own funds" of universities included in "other national"
cluded		"Own funds" of universities included in "Government"
cluded		
cluded		
cluded		
cluded		"Other national" includes PNP only. The "own funds" of Higher Ed. are included in "Government"
cluded		In 1972 "other national" includes funds from Private Non-Profit only (less than 0.5% of GERD)
cluded	*This distribution has varied. ISY : 1964 = 81% 1964 + 19% 1965 1967 = 11% 1966 + 89% 1967 1969 = 12% 1968 + 88% 1969 1971 = 6% 1970 + 82% 1971 +12% 1972 1973 = 85% 1973 + 15% 1974	
cluded	*This distribution varies slightly from year to year	
cludes social ciences and sychology, cept in the siness Enterise sector.	Not GERD but total current expenditure only, including depreciation in the Business Enterprise sector	

) These % were used to reweight the GDP data and the GDP deflators.

See Column 5.

Sweden :

1964-1971 : National responses to the OECD ISY's.

1973 : Preliminary data for the 1973 ISY provided by the national authorities for the present document.

United Kingdom :

1964-65 - 1969/70 : United Kingdom response to 1971 ISY Table T6.

1972/73 : OECD estimate based on "Industrial Expenditure on Scientific R&D : Provisional Results of the 1972 Survey". Trade and Industry September 5th 1974 and "Research and Development Expenditures", CSO, 1973.

United States :

1964-1970 : "National Patterns of R&D Resources 1953-74"

1971-1974 : "National Patterns of R&D Resources 1953-75".

II. TOTAL GOVERNMENT R&D FUNDING AND ITS DISTRIBUTION BY OBJECTIVE

A. Total Government R&D Funding

Total government R&D funding covers all current and capital expenditures by government (whether at home or abroad) as reported by the funding agency. R&D in the social sciences and humanities is generally included, as is the support of R&D performed by universities financed out of their general grant from the ministry of education.

R&D Data Quoted in Graph IV and Graph V and Table 5

1. Coverage : See tabulation B.
2. Sources :

a) EEC countries not elsewhere specified

"Public Expenditure on Research and Development in the Community Countries : Analysis by Objectives, 1973-74". Second Report from the Sub-Committee "Statistics" to the Committee on Scientific and Technical Research" (CREST).

b) Other countries

Austria :

1975 Report on Research by the Federal Government to Parliament (preliminary data) submitted for the present report.

Belgium :

"Analyse par objectifs du programme budgétaire de politique scientifique" 1965-1974 (budget ajusté) et 1975 (déposé). Data submitted by the Belgian authorities, based on this publication.

Canada :

"Federal Government Activities in the Natural Sciences, 1972-74", p. 67, Table 2. Statistics Canada.

Finland :

Data drawn from a national science budget analysis submitted for the present report.

France :

Data submitted by the national authorities (DGRST) for the present report.

Iceland :

Rannsóknastarfsemi á Íslandi á Svidi Raunvísinda 1972 og 1973 Reykjavik, January 1975).

Italy :

1963-1970 : Changing Priorities for Government R&D. An Experimental Study of Trends in the Objectives of Government R&D in 12 OECD Member countries 1961-72, OECD, 1975.

1971-74 : EEC, *op. cit.*

Japan :

Government intramural expenditure - "Changing Priorities for Government R&D", updated by the national authorities for 1970-74. Public funding of Higher Education : OECD estimates.

Netherlands :

Graphs IV & V : "Changing Priorities for Government R&D", updated by the Secretariat using the Netherlands science budget (Table 5) and EEC, *op. cit.*

Norway :

Graphs IV and V : "Changing Priorities for Government R&D", updated by the Secretariat using budget documents. Table 5 : See Sources and Methods III.

Portugal :

Data submitted by the national authorities for the present report.

Sweden :

"Changing Priorities for Government R&D", updated by the Secretariat using budget documents.

United Kingdom :

"Changing Priorities for Government R&D", updated by the Secretariat using the table on R&D expenditures included annually in "Memorandum by the Chief Secretary to the Treasury" ; and the EEC report, *op. cit.*

United States :

1964-70 : "National Patterns of R&D Resources 1953-74"

1971-74 : "National Patterns of R&D Resources 1953-75"

B. Breakdown of Government R&D Funding by Objectives

The distribution by objective is usually made according to the principal aim of the funding agency when committing the funds to R&D. As this is a relatively new type of analysis, national and international classifications vary somewhat. As a result of recent discussions the three main international classifications, those of the OECD, the EEC and Nordforsk will, in the future, be compatible.

1. General remarks on trends in the OECD area in the 1960's

These are based on Changing Priorities for Government R&D. An Experimental Study of Trends in the Objectives of Government R&D Funding in 12 OECD Member countries, 1961-1972 , OECD, 1975.

2. Sources of data backing the text on defence, energy and the social objectives

a) EEC countries, not elsewhere specified (1969-74)
"Public Expenditures on Research and Development in the Community Countries" EEC, op. cit.

b) Other countries

Belgium :

"L'analyse par objectifs du programme budgétaire de politique scientifique" 1965-1974 (budget ajusté) 1975 (déposé).

Finland :

Budget data provided by the national authorities.

France : (Table 7)

"L'enveloppe-recherche en 1975", Le Progrès scientifique, N° 173, Nov.-Dec. 1974.

Norway :

Updated series prepared by the Secretariat for the publication Changing Priorities, op. cit. and "Bevilgninger over Statsbudsjettet til forsknings - og utviklingsarbeid 1975".

Sweden :

Updated series prepared by the Secretariat for Changing Priorities, op. cit.

United States :

(Table 6) "An Analysis of Federal R&D Funding by Function, 1969-75" (NSF 74-313).

Tabulation B

GOVERNMENT R&D FUNDING

Country	Years	Type Expenditure Data	Soc. Sciences & Humanities
Austria	1970-75	Current & capital	Included
Belgium	1965-75	Current exp. only, domestic & abroad	Included
Canada	1964/65-1973/74 (1.4-31.3)	Current & capital	Excluded
Denmark	1969-74	Current & capital, domestic & abroad	Included
Finland	1970-75	Current & capital	Included
France	1966-73	Current & capital, domestic & abroad	Included
Germany	1961-74	Current & capital domestic & abroad	Included
Iceland	1971-73	Current & capital	Excluded
Ireland	1969-73	Current & capital, domestic & abroad	Included
Italy	1963-74	Current & capital domestic & abroad	Included
Japan	1961/62-72/73 (1.4-31.3)	Current & capital, intramural + public funding of Higher Education	Included
Netherlands	1961-75	Current & capital, domestic & abroad	Included
Norway	1961-75	Current & capital	Included
Portugal	1971+1972	Details not avail.	Included
Sweden	1961/62-75/76 (1.7-30.6)	Current & capital	Included
United Kingdom	1961/62-74/75 (1.4-31.3)	Net central gov't. exp. on R&D, current & cap., domestic & abroad	Partially inc. as from 1966/67 ; excluded for prev. years.
United States	1961/62-74/75 (1.7-30.6)	Current exp. only	Soc. sciences & psychology included ; humanities excluded

C. Coverage of Social Objectives

The following were considered as "socially oriented" objectives in the countries studied :

Countries	Objectives
Denmark France Germany Ireland Italy Netherlands United Kingdom	The following EEC objectives : "Protection and Promotion of Human Health" (including "R&D on noxious phenomena") and "Promotion of Research in the Social Sciences and Humanities".

Belgium :

National objectives :
"Research Towards Progress in the Art of Healing"
"Research Towards Progress in the Human Environment" (partial education, social relations and progress of law) and Pollution (part of the objective Knowledge and Conservation of the Environment).

United States :

National "functions" :
Health
Environment
Education
Income Security and Social Services
Crime Prevention and Control

Norway and Sweden :

Objectives drawn from the OECD Experimental Study *Changing Priorities for Government R&D*, *op. cit.* :
Health
Pollution
Public Welfare
Other Community Services.

III. NON R&D DATA

A. Deflators (Graph I, Table 1 and Graph IV)

Since it was not possible to calculate special "R&D deflators", the implicit Gross Domestic Product deflator has been used. This only approximates the real change in R&D prices. The resulting trends shown in Graphs I and IV are broadly correct but the R&D growth rates quoted in Table 1 should be used for broad comparisons only.

The deflators were derived from a series of GDP data supplied by the Economic Statistics and National Accounts Division. Those for 1974 and 1975 are based on their estimates for these years. ("OECD Economic Outlook - 1974"). The deflators have been adjusted to cover the same time period as the R&D data. In some cases this differs between Graph I (GERD) and Graph IV (total government R&D funding).

B. 1961 Dollars (Graphs I and IV)

The national currency data were deflated and then converted into US dollars using the 1961 exchange rate. It should be noted that this procedure reveals trends over time but gives very different results for the later years than would a straight conversion into dollars at say 1971 exchange rates. Graphs I and IV should thus not be used to make international comparisons of the level of GERD or of total government R&D funding at the end of the period.

C. Gross Domestic Product (GDP) (Graph II, Table 3B)

In the past the percentage of Gross National Product (GNP) devoted to R&D has been a well-known science policy indicator. Following a revision of the System of National Accounts, the OECD no longer publishes data on GNP. This has been replaced by the Gross Domestic Product (GDP). In line with OECD practice this measure of national wealth has been adopted for this report. GDP data was obtained from the Economic Statistics and National Accounts Division. They have been reweighted to cover the same time period as the GERD data in each country. For both reasons the resulting percentages may be slightly different from those known to national science policy makers. The trends are the same.

D. Total Government Expenditures (Graph V)

This series is the sum of Government Final Consumption expenditure and of Government Gross Fixed Capital Investment taken from "National Accounts of OECD countries", 1961-1972 and 1962-1973 - Vol.I (except Belgium). For those countries where total government R&D funding excludes capital expenditures these have also been excluded from the denominator. The data have been reweighted to cover the same period as the total government R&D funding data.

E. Total Government Expenditures (Table 5)

The measures of total government expenditures used in this table are consistent over time but not between countries. The sources for the data and other coverage are as follows.

1. EEC countries not elsewhere specified

R&D as a % of "Budget global de l'Administration centrale" as quoted in "Le Financement public de la recherche et du développement dans les pays de la Communauté 1973-74".

2. Other countries

Austria :

"Bundeshaushalt, 1973-1975" supplied by the Austrian Delegation to OECD. The R&D data are for federal government only.

Belgium :

Exceptionally, national data on government final consumption expenditure were used in Graph V. They were supplied by the Belgian authorities for 1965, 1967, 1969, 1971-1975. In Table 5 total budget data provided by the national authorities were used.

Finland :

Data were provided by the national authorities on government R&D funding as a percentage of government total consumption expenditure.

Ireland :

Net estimates (i.e. Gross estimates less appropriations-in-aid). Supplied by the national authorities.

Norway :

Total government budget data and the share of R&D as quoted in NAVF, NLVF, NTNF "Bevilgninger over statsbudsjettet til forsknings og utviklingsarbeid 1975". These R&D data are not entirely comparable with those quoted in Graph IV and Graph V.

Sweden :

Total budget expenditures quoted in "The Swedish Budget 1975-76" and previous annual editions "Summaries in English published by the Ministry of Finance).

United States :

Total budget outlays provided by the national authorities.

CONCLUSION

The Ministers noted that the contribution of science and technology to the solution of the problems of modern society, and to the development of the economies of their countries, as well as to the world economy, has been a key element and will become of even greater importance. The scientific and technological resources of the OECD countries, which account for over two-thirds of the world's research capacity, are a determining factor in their ability to achieve the individual and collective aspirations of their peoples and to help solve the problems facing the developing countries, including developing Member countries. The vigorous research and development programmes pursued by the OECD countries serve to improve the understanding of the problems affecting the standards of living and the quality of life in such fields as energy, nutrition, environment and welfare, and to provide means for dealing with them.

Ministers recognised that in order to succeed in these aims, scientific and technical knowledge must be more effectively applied to the definition, implementation and content of Government policies at all levels. This is all the more important in view of the complexity of problems facing today's society.

The overall theme of the meeting was thus Science and Technology in the Management of Complex Problems. In this context, Ministers assessed the new challenges resulting from the major economic, political and social changes since their last meeting in 1971. They reviewed current problems facing the research system of Member countries, in particular, their social and management aspects, the mutual responsibilities of both policy makers and scientists, and the increased perception on the part of the public of the implications of scientific and technical advance. They also considered the contributions which science and technology could make to a better social equilibrium in the world, including the issues discussed at the recent Meeting of the OECD Council at Ministerial level : commodities and relations with developing countries. They emphasised that in all these areas the OECD's Committee for Scientific and Technological Policy had an important role. They also expressed appreciation of the current work of the Committee, particularly in relation to the application of science and technology to the solution of social problems.

The discussions centred on the following six topics, each of which has important national and international implications :

I. THE ROLE OF SCIENCE AND TECHNOLOGY IN POLICY MAKING

Given the complexity of the problems facing contemporary society and the need for medium-and long-term prospective planning to ensure the compatibility of options and to anticipate future problems, ministers stressed the importance of bringing scientific insight to bear on general and sectoral policies of governments. They recognised the need for the more effective use of science and technology in the preparation and implementation of government policies.

To this end, they saw the need to :

- assess the experience of Member countries of the ways in which science and technology are used in shaping government policies,
- identify the elements of broad prospective analysis required for medium-and long-range planning, with a view to achieving a better understanding of the relationship between social, economic, and scientific and technological factors,
- promote the integration of social sciences in the policy-making process.

II. THE MANAGEMENT OF THE RESEARCH SYSTEM

The Ministers recognised that the research system cannot be geared solely to the short-term problems of society and emphasised the need to ensure that the overall capacity of the system should be sufficiently flexible to accommodate research on both short-and long-term problems. The levelling-off of resources allocated to research and development, observed in several Member countries, makes it difficult to ensure this flexibility, which calls for examination of the problems involved, especially as far as research structures and mobility of scientists are concerned.

The Ministers agreed that efforts should be made, particularly at national level, to provide a better basis for decision-making by strengthening capabilities for the analysis and interpretation of research and development data. They further recommended that OECD should examine methods to increase the efficiency of the research system, and the management problems which arise from changes in levels of research and development funding.

III. SCIENCE, TECHNOLOGY AND SOCIETY

The Ministers recognised the need to ensure that the increasing emphasis on research directed towards social objectives leads to a better application of science and technology towards social needs in areas such as environmental protection, mass transportation, urban planning and the humanisation of working conditions. They also agreed that policies for the stimulation of technology must take into account possible undesirable social effects.

The Ministers emphasised the importance of promoting a closer association of the natural and social sciences, particularly in addressing problems requiring a broad interdisciplinary approach and perspective. They pointed to the need for better understanding of the factors involved in the conduct of inter-disciplinary research aimed at solving social problems. In this connection, since the social sciences have an essential role in the understanding and management of the problems of society, Ministers expressed the wish that OECD would intensify its work on the development and utilisation of these sciences in relation to policy-making.

IV. THE NEED FOR PUBLIC INVOLVEMENT

Ministers recognised the need to establish a broad and open dialogue between the public and those responsible for scientific and technological activities. Public understanding of new technologies has become an essential factor in their acceptance and therefore their utilisation. Ministers expressed the wish that the OECD should review the experiences of Member countries regarding public participation in science and technology. It was hoped that this initiative would materially assist Member Governments to devise effective means of informing the public of the implications of new technological developments, soliciting their reactions, and engaging them in the decision-making process.

V. A NEW VISION OF NATURAL RESOURCES

Ministers emphasized the urgent need for a better husbanding of the world's natural resources, which has become a critical issue. They recognised the complex nature of the problem which should be the subject of detailed examination. Such an examination should take into account all the interactions involved and should lead to a better understanding of the choices available and of their implications, including possible changes in demand for raw materials based on the potential impact of improved utilisation and new technological developments.

Ministers expressed the wish that OECD, acknowledging that the nature of the issues at stake requires the involvement of other international organisations, at both regional and world level, would encourage studies in this direction. Furthermore, they underlined the strong interdependence between industrialised and developing countries and the consequent need to establish mutual cooperative efforts in this field bearing in mind the global perspective.

In connection with the critical problem of food resources, they noted with interest the proposal by the United States on the need to promote food research and development. They invited the Secretary-General to prepare specific proposals in this area in consultation with the OECD Committees concerned, and taking account of on-going work in other international organisations, notably FAO, and in the light of the deliberations of the forthcoming OECD meeting of Directors of Agricultural Research.

VI. INTERNATIONAL CO-OPERATION

Acknowledging that OECD countries carried out the major proportion of science and technology activities and taking into account the growing complexity of the problems facing society, Ministers stressed the need to adapt current arrangements to ensure better international co-operation in science and technology, notably on the basis of bilateral and multilateral arrangements. They considered that greater emphasis should be placed on co-operation between Member countries in interdisciplinary fields.

Conscious of the special responsibilities of their countries toward the developing nations, Ministers agreed on the importance of developing a broader perspective in the conduct of their national research and development programmes. They recognised that a new pattern of international relations was developing, based on interdependency, in which exchanges mutually beneficial to all countries could be achieved. Ministers expected that the forthcoming 7th Special Session of the General Assembly of the United Nations would offer a further opportunity to discuss problems of science and technology in a global context.

Ministers agreed on the necessity to actively reinforce co-operation between their countries and the developing world, with a view to facilitating the transfer of, and access to, appropriate technologies in order to satisfy the needs of developing countries more effectively and to strengthen the scientific and technological capabilities of these countries. They therefore invited OECD to pay special attention to the various aspects of this problem, including the training of qualified scientific manpower.

OECD SALES AGENTS
DEPOSITAIRES DES PUBLICATIONS DE L'OCDE

OECD PUBLICATIONS, 2, rue André-Pascal, 75775 Paris Cedex 16 - No. 35.987 1976

PRINTED IN FRANCE